Cognitive Systems Monographs

Volume 14

Bojan Jakimovski

Biologically Inspired Approaches for Locomotion, Anomaly Detection and Reconfiguration for Walking Robots

Rüdiger Dillmann, University of Karlsruhe, Faculty of Informatics, Institute of Anthropomatics, Humanoids and Intelligence Systems Laboratories, Kaiserstr. 12, 76131 Karlsruhe, Germany

Yoshihiko Nakamura, Tokyo University Fac. Engineering, Dept. Mechano-Informatics, 7-3-1 Hongo, Bukyo-ku Tokyo, 113-8656, Japan

Stefan Schaal, University of Southern California, Department Computer Science, Computational Learning & Motor Control Lab., Los Angeles, CA 90089-2905, USA

David Vernon, Khalifa University Department of Computer Engineering, PO Box 573, Sharjah, United Arab Emirates

Author

Dr.-Ing. Bojan Jakimovski
Bionics4Robotics
Postfach 900609,
81506 München, Germany
E-mail: contact@bionics4robotics.com

ISBN 978-3-642-22504-8 e-ISBN 978-3-642-22505-5

DOI 10.1007/978-3-642-22505-5

Cognitive Systems Monographs ISSN 1867-4925

Library of Congress Control Number: 2011934501

Typeset & Cover Design: Scientific Publishing Services Pvt. Ltd., Chennai, India.

Printed in acid-free paper

5 4 3 2 1 0

springer.com

Abstract

The increasing presence of mobile robots in our everyday lives introduces the requirements for their intelligent and autonomous features. Therefore the next generation of mobile robots should be more self-capable, in respect to: increasing of their functionality in unforeseen situations, decreasing of the human involvement in their everyday operations and their maintenance; being robust; fault tolerant and reliable in their operation.

Although mobile robotic systems have been a topic of research for decades and aside the technology improvements nowadays, the subject on how to program and making them more autonomous in their operations is still an open field for research.

The classical formal methodologies which have been "dominating" the robotics segment for long time start to prove that they are perhaps not adequate to cope with the increasing complexity of the robotic systems. Many different research directions have been considered on how to overcome these problems.

Applying bio-inspired, organic approaches in robotics domain is one of the methodologies that are considered that would help on making the robots more autonomous and self-capable, i.e. having properties such as: self-reconfiguration, self-adaptation, self-optimization, etc.

In this book several novel biologically inspired approaches for walking robots (multi-legged and humanoid) domain are introduced and elaborated.

They are related to self-organized and self-stabilized robot walking, anomaly detection within robot systems using self-adaptation, and mitigating the faulty robot conditions by self-reconfiguration of a multi-legged walking robot. The approaches presented have been practically evaluated in various test scenarios, the results from the experiments are discussed in details and their practical usefulness is validated.

Contents

Chapter 1
Introduction

Robotic systems nowadays are getting increasingly complex in their design and implementation. In order to fulfill the proposed requirements, systems often consist of many software and hardware units, realizing various functionalities and cooperating together. Declaring them as autonomous means that they should have the ability to dynamically adjust and execute their tasks without human intervention. Additionally, they should be reliable and also tolerant to various system malfunctions.

Future robotic systems should be able to demonstrate self-x properties such as self-organization, self-reconfiguration, self-healing and the like. Having these kinds of properties, robotic systems would be able to demonstrate their autonomic property, and this will also aid in shortening their development and maintenance time.

However, the complexity of the classical approaches for robot system modeling has introduced a need for developing and applying new concepts and methodologies towards creating self-capable, more robust, and dependable systems. In order to achieve this, engineers have used different biologically and organically inspired approaches. For example, some of the algorithms for motor control and walking gait pattern generations for the domain of joint leg walking robots have been inspired and developed on observations seen within animals and functioning of the neural circuitry. In that context, research has been done and is presented in this book on introducing novel biologically inspired approaches with their practical applications for domain of *self-organizing, self-optimizing, and self-reconfiguring* walking robots. Joint leg walking robots and simple walking robots use their legs for their movement over a terrain. Depending on how many legs they have, they can belong to one of many categories, from two legged humanoid robots up to many legged robots having three, four, six, or eight legs, where each leg can be built out of several joints/segments. Walking robots can have many joints and therefore many degrees of freedom (DOF). However, having many degrees of freedom introduces difficulties in developing appropriate methods for controlling such complex walking robots, monitoring their health status, making them fault-tolerant, etc.

The research presented in this book tries to give an answer on some of the open questions for fault-tolerant robotics domain. This includes self-organized,

B. Jakimovski: Biologically Inspired Approaches, COSMOS 14, pp. 1–3.
springerlink.com

self-stabilized robot walking and self-adapting methods for failure detection and self-reconfiguration after a failure has been detected within the robotic system, so the robotic system can still continue with its mission despite the faulty conditions within itself.

Research presented in this book includes:

- biologically inspired robot control architecture;
- self-organizing hexapod robot walking;
- biologically inspired self-stabilizing humanoid robot walking;
- biologically inspired self-adapting approach for anomaly detection by a robot;
- biologically inspired approach for self-reconfiguration by a hexapod robot.

The structure of the thesis is organized as follows:

- In the 2nd chapter, a general overview is given on biologically inspired computing and self-x properties including short introduction to terms related to: bionics, organic computing, autonomic computing, self-x properties, and emergence;
- The 3rd chapter is about joint leg walking and hybrid robots, with a small review about state of the art of hexapod robots. In this chapter detailed descriptions about the hexapod robot demonstrators OSCAR-2, OSCAR-3, and OSCAR-X are given. State of the art humanoid robots are further introduced and also a detailed description of the S2-HuRo humanoid robot demonstrator is given.
- The 4th chapter describes notions about a biologically inspired robot control architecture. An overview is given on commonly used robot control architectures such as: reactive, subsumption, deliberative, and hybrid control. After the introduction on commonly used robot control architectures, an introduction is given on self-organizing robot architectures such as: autonomic control architecture, Organic Robot Control Architecture (ORCA), and their characteristics. The distributed ORCA which is related to the research experiments is explained in detail and new ideas are also given about an enhanced "stem" based ORCA architecture.
- The 5th chapter first gives an introduction on locomotion seen by insects and animals and Central Pattern Generator (CPG) for walking pattern generation. Then a concept for self-organizing emergent robot walking gait with distributed pressure on robot's feet is explained. Further explained in this chapter is the firefly-inspired synchronization of a robot's walking gait. Included are the experiments done on prolongation and shortening of the robot's swing and stance phases using firefly-inspired synchronization.
- In the 6th chapter a biologically inspired approach for optimizing the walking gait of a humanoid robot is explained. Symbiosis as a biologically inspired approach for self-stabilization of humanoid robot walking gait is elaborated and details about the SelSta approach and its main parts are

discussed. The chapter concludes with the results of experiments done on the SelSta approach and the usefulness of the SelSta approach.

- The 7th chapter is about biologically inspired approaches for failure detection within a robotic system. First it gives an overview on the approaches for robot fault / anomaly detection, followed by an introduction on Artificial Immune Systems. Then it introduces the Artificial Immune System (AIS) based Robot Anomaly Detection Engine (RADE) approach. After RADE is explained in more detail, results from the experiments done with the AIS-inspired RADE approach are presented and discussed.
- The 8th chapter introduces an approach for robot self-reconfiguration of a hexapod robot system based on an biological inspiration - swarm intelligence. An overview of swarm intelligence, flocking behavior, and boids is given first. Then S.I.R.R., a Swarm intelligence based approach for robot reconfiguration, is introduced and explained in detail. After the introduction of the S.I.R.R. approach, results from simulation of S.I.R.R.-based hexapod robot reconfiguration are presented. The results from real robot reconfiguration experiments done with the S.I.R.R. approach on the hexapod robot OSCAR-2 are then presented for validation of the simulation experiment's results. Chapter 8 ends with the results from real robot reconfiguration experiments done with the S.I.R.R. approach, presentation of leg amputations on robot OSCAR-X, and an explanation of practical usefulness of S.I.R.R.
- The 9th chapter gives a conclusion on the research presented in this book and the importance of the biologically-inspired approaches introduced for the walking and general robotics domain.

Chapter 2
Biologically Inspired Computing and Self-x Properties

2.1 Bionics

Bionics is the application of biological methods and systems found in nature to the study and design of engineering systems and modern technology [Bio10].

The term Bionics (from biology and electronics) is sometimes interchangeably used for a Biomimetics and Biomimicry (from bios = life, and mimesis = to imitate). Bionics is related to applying ideas seen in nature for solving scientific, technical, or engineering problems. Biomimetics is therefore an interdisciplinary field where scientists from different scientific fields like Biology, Physics, Chemistry, and Engineering work together towards developing various solutions based on observations of processes seen in nature. All these techniques are based on solving new problems from solutions to previous problems found in nature.

For example, such biologically inspired concepts can be related to various organizational building principles seen within bacteria, flora, fauna, etc.

There are many innovations and products developed that can be mentioned as examples for practical usefulness of biologically inspired concepts: self-assembling glass inspired by sea sponges; bacterial control inspired by red algae; solar cells inspired by leaves; friction-free fans inspired by nautilus; building material from CO_2 inspired by mollusks; self-cleaning surfaces inspired by lotus plant [Nat10], etc.

Besides the natural assembling techniques which can be useful for building novel materials with new and perhaps superior properties than that of the current existing materials, the social and organizational principles in nature like feeding, mating, foraging, and swarming are also found to be useful for engineering domain.

Bionics is more related to implementing the approach found in nature as an idea instead of imitating the biologically structure behind it, which is more closely representing the Biomimetics.

Nowadays there is an increasing trend of applying and adopting biologically inspired approaches for the general domain of computer science as well as for the domain of robotics.

B. Jakimovski: Biologically Inspired Approaches, COSMOS 14, pp. 5–7.
springerlink.com

More and more of the approaches implemented in the field of computer engineering are related to artificial intelligence and optimization, such as artificial neural networks, genetic algorithms, and ant optimization algorithms.

2.2 Organic Computing

Organic computing initiative [Mül04] [Sch05], is related to development of technical systems that act autonomously and dynamically adapt to the environment. Those systems must exhibit lifelike properties and function in independent way. That is why they are called "organic" or "organic computing systems". At the same time they should be also robust, safe, and trustworthy.

Therefore, organic computing is a type of biologically inspired computing that attempts to develop approaches for technical systems exhibiting self-x properties such as self-organization, self-configuration, self-optimization, self-healing, and self-explaining.

The research presented in this book is directly associated with Organic Computing and developing self-x biologically inspired approaches that would enable the robotic systems to act in more independent and autonomous ways.

2.3 Autonomic Computing

The Autonomic computing initiative was proposed in the IBM [IBM01] manifesto and states the need for development of autonomic IT systems that would overcome the ever growing complexity of current IT systems. The main requirement for such autonomic systems would be that they are self-manageable and also capable of providing reliable services and minimizing the human administrator intervention and thus minimizing the probability of human errors.

Autonomic systems are therefore systems that can manage themselves without human intervention. They must be capable of incessant autonomous work given only high-level objectives from the administrators [KeC03]. In order to achieve the autonomic system's property, such systems must have the self-x properties self-reconfiguration, self-organization, and self-healing.

2.4 Self-x Properties

Self-x properties are closely related to biological processes found in nature, namely the processes in nature that can be often seen as self-organizing, self-optimizing, and self-healing processes.

Such self-x properties have been proven useful when translated to scientific domains: techniques for new materials development, engineering, new approaches for the IT industry, etc.

Approaches mentioned in this book are mainly related to the development of various algorithms for joint leg walking robots domain that enable the robots to exhibit the so-called self-x properties for various circumstances such as robot self-reconfiguration, robot walking gait self-optimization, and robot self-healing.

General definitions for the terms Self-Configuration, Self-Optimization, Self-Healing, and Self-Protection can be found by Autonomic Computing [KeC03] and also can be used interchangeably for other technical domains, as well as for the robotics domain.

Self-Configuration: Automated configuration of components and systems follows high-level policies. The rest of the system adjusts automatically and seamlessly.

Self-Optimization: Components and systems continually seek opportunities to improve their own performance and efficiency.

Self-Healing: The system automatically detects, diagnoses, and repairs localized software and hardware problems.

Self-Protection: The system uses early warning to anticipate and prevent system-wide failures.

2.5 Emergence

Emergence is one of the phenomena often observed in nature. Emergence can be defined as "the arising of novel and coherent structures, patterns, and properties during the process of self-organization in complex systems" [Gol10]. The emergence can sometimes be summarized as: Whole is more the than sum of its parts. This means that in systems exhibiting emergence, the behavior or the property of the whole system cannot be deducted from the properties of individual components composing that system. Such a definition is closer to the view of "strong emergence". On the other side, "weak emergence" is related to the emergence that is traceable, i.e. the emergent property can be reduced to the property of individual components.

For emergence is often said to be a "bottom-up" process.

There are many examples of emergent processes that can be seen in nature or in biological systems. For example: the sand dunes, water waves, swarming schools of fish, flocking of birds, slime molds, ant colonies' self-sustainability, etc.

In complex systems where safety is not a critical issue (since the completely safe system's behavior cannot be guaranteed), emergence is sometimes used to lower the effort of developing the needed system's functionality.

Emergence is also popular for domain of robotic systems. Examples include: emergence of gait patterns for robot walking [AWY99], emergence of communication within multi robot systems [Lip07], emergent behaviors of autonomous robots [AnD90], etc.

The practical usefulness of the emergence concepts is explained in further chapters, applied to the research on the hexapod robot OSCAR, more precisely for its walking gait generation.

Chapter 3
Joint Leg Walking and Hybrid Robot Demonstrators

3.1 Introduction

Design of two and multi-legged robots like four legged, six legged, and eight legged robots shows a practical usefulness of Bionics for the domain of robotics. Depending on the number of legs, such robotic platforms can be inspired from body constitution, walking mechanics, and behavior of humans, animals (four legged), insects (six legged), or spiders (eight legged). There is also another kind of robotic designs depicted as hybrid robots, which have a mixture of concepts seen in nature and artificial designs. These robots can be often found having tracks, other special leg designs, wheeled-legged robot designs and the like.

In this chapter several joint leg walking and hybrid robots are discussed. The "joint leg" term is related to robots that have legs built out of servos, representing their joints. The term "hybrid" found by hybrid robots is related to robots that have both legs built out of servo joints and wheels attached to their legs.

Robots demonstrators presented in this chapter have been used as demonstrators for biologically inspired approaches and algorithms researched and elaborated in this book. These robots can be categorized into three types: humanoid robots, hexapod-robots, and hybrid wheeled-legged robots.

3.2 Hexapod Robots

Hexapod robots belong to the group of joint leg walking robots having six legs where the legs are consisting of multiple servo joints. The legs of the robot are usually symmetrically distributed in two different groups spatially located on the two opposite sides of the robot's body. The design of hexapod robots is often inspired by locomotion systems seen in insects like cockroaches, stick insects, and the like.

In comparison with the four legged walking robots or quadruped robots, hexapod robots have intrinsically more redundancy due to the higher number of the legs and thus can be theoretically more flexible over uneven terrain. Hexapod

B. Jakimovski: Biologically Inspired Approaches, COSMOS 14, pp. 9–21.
springerlink.com

robots differ from robots that have "native" spider-like biomimetic design having eight legs distributed on the two sides of the robot's body. Although the eight legged robots may have higher degree of redundancy and perhaps provide better agility for the robot over rugged terrain, they also need more energy for their functioning, which in turn affects the size and mobility of the robot.

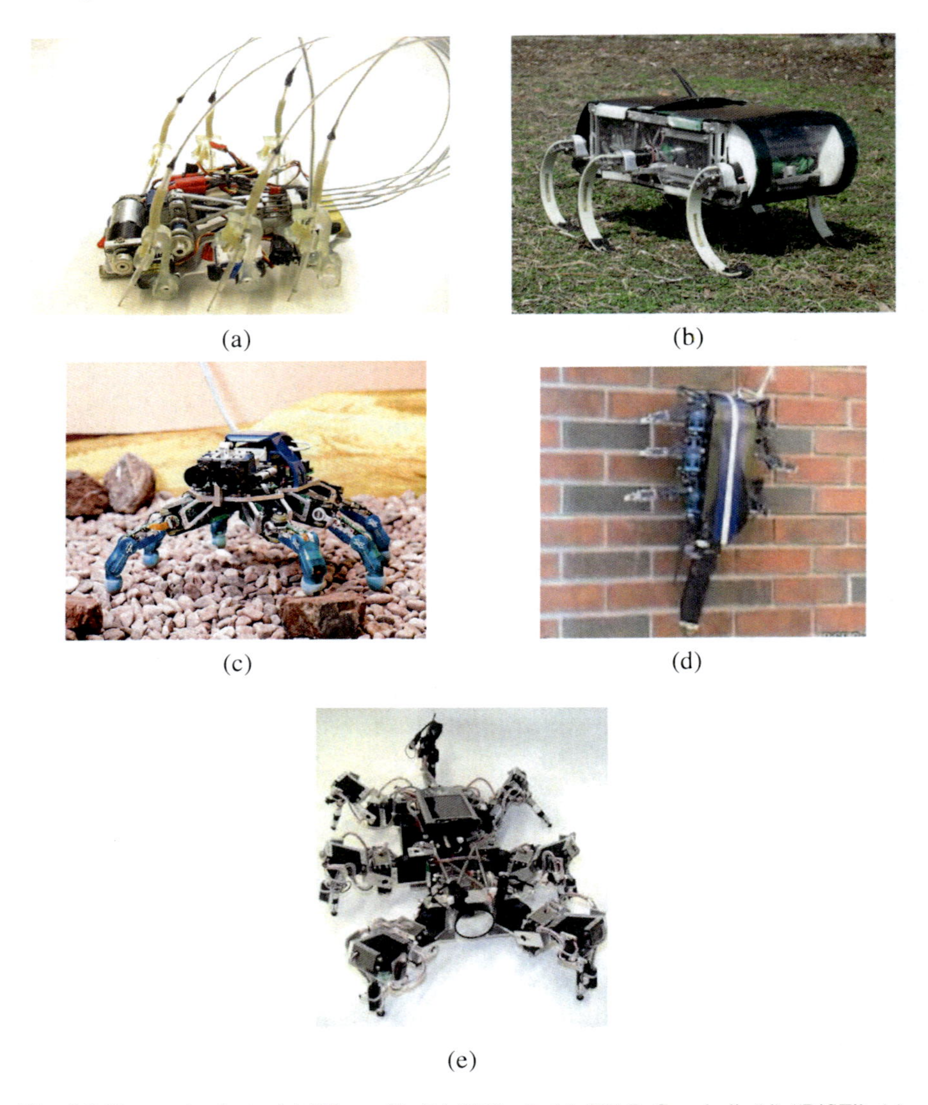

Fig. 3.1 Hexapod robots: (a) "iSprawl"; (b) "RHex"; (c) "DLR Crawler"; (d) "RiSE"; (e) "AMOS-WD06".

3.2.1 State of the Art – Hexapod Robots

There have been many types of hexapod robots that have been used for demonstration purposes in the research on biologically inspired locomotion. Some of the current (Feb, 2010) state of the art hexapod robots include: "iSprawl" [KCC06], "RHex" [AMK01], "DLR Crawler" [GWH09], "RiSE" robot [SGF06], "AMOS-WD06" [STW10] (Figure 3.1). However, the mentioned hexapod robots all differ in the technology that they use for their locomotion. For example, their leg design differs from one robot to the other and the moving concept of the joints by the legs also differs.

Movement of the legs for the "iSprawl" robot is periodic, generated by push-pull actions using flexible cables and servo motors [KCC06]. For the "RHex" robot, the movement of the legs is related only to the rotary motion of the legs [AMK01]. The "DLR Crawler" design was based on the "DLR-HAND II" [BFH03], therefore the joint based fingers of the "DLR-HAND II" are adapted to serve as legs for the "DLR Crawler" [GWH09]. For the "RiSE" robot, the legs are moved by two electrical actuators per leg, with biologically inspired adhesive structures located on the feet, which enable the "RiSE" robot to climb on vertical wall surfaces, trees, etc [SCM06]. The "AMOS-WD06" robot implements legs consisting of three servos each, resulting in eighteen servos for locomotion of the hexapod robot [STW10].

By the presented "state-of-the-art" robots there are improvements that can be seen in comparison with some older hexapod robot designs, however less has been done on introducing fault tolerant mechanisms within the robots itself, which will give the robots the robustness to be functional also in situations when they experience some malfunctions within their components, by means of reconfiguring the body and leg postures.

And this is the key difference in comparison to the robot demonstrators OSCAR, in particular OSCAR-X, which were developed at Institute für Technische Informatik, University Lübeck and described in the following sub-chapter.

3.2.2 Hexapod Robot Demonstrator – OSCAR (Organic Self Configuring and Adapting Robot)

OSCAR (Organic Self Configuring and Adapting Robot) is a six legged walking robot, used as a demonstrator for testing some of the newly-developed biologically inspired approaches and algorithms presented in this book.

OSCAR represents the series of built hexapod robot demonstrators used in the interdisciplinary research, where the legs are distributed spatially in a circle on the robot's body. Most of the robots in OSCAR series have integrated an on-board embedded system, sensors (ultrasonic, infrared, acceleration, and inclination sensors), and actuators (analog and digital servos). Such a typical spatial distribution of the robot's legs in a circle is represented in (Figure 3.2).

Fig. 3.2 Spatial distribution in circle of legs by OSCAR series of robots.

The OSCAR series consists of the following robots: OSCAR-1, OSCAR-2, OSCAR-3, and OSCAR-X, which are described below. The first two robots, OSCAR-1 and OSCAR-2, were mostly based on the Lynxmotion® robot kit - "AH3-R (18 Servo Walker)" with 18 degrees of freedom (DOF), Hitec HS-645 Servos and aluminum based leg design [Lyn06].

3.2.2.1 Hexapod Robot Demonstrator – OSCAR - 1

Hexapod robot OSCAR-1, built in year 2006 is the first in the series of OSCAR robots. Its hardware is based on the "AH3-R (18 Servo Walker)" robot kit with six legs distributed spatially in a circle and additional on-board electronics such as "JControl" (a Java based embedded system for robot control), servo controller SD-21, Hitec analog servos HS-645, binary contact sensors on the robot's feet, and NiMH batteries (Figure 3.3). Each leg by the robot is made up of three servos. There are also integrated ultrasonic sensors on three of the robot's legs.

Fig. 3.3 Hexapod robot OSCAR-1

3.2.2.2 Hexapod Robot Demonstrator – OSCAR - 2

OSCAR-2 (Figure 3.4) is the second in the series of OSCAR robots, similarly built as OSCAR-1.

The OSCAR-2 in comparison to the OSCAR-1, has the following modifications:

- pressure sensors (Figure 3.5);
- 18 modified HiTec HS-645 servos (Figure 3.6);

The modified servos provide feedback for the level of servo current, so the torque can be monitored while the robot is walking. The modification is clearly visible due to the number of wires that come out from the servos, namely that the wires are directly connected to the potentiometer output reading pins inside of the servos.

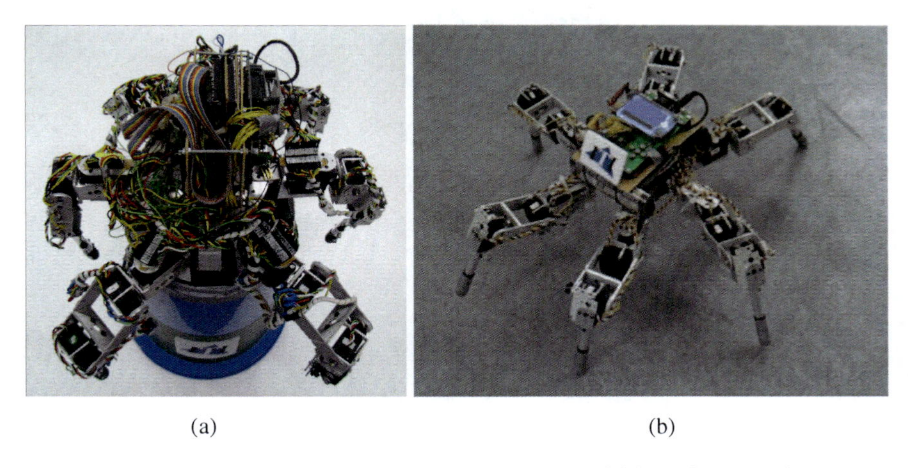

(a) (b)

Fig. 3.4 Hexapod robot OSCAR-2. (a) Experimental robot OSCAR-2 setup - from above; (b) Robot OSCAR-2 in movement.

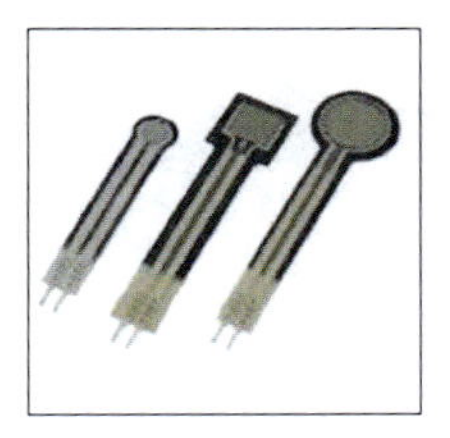

Fig. 3.5 Pressure sensors type FSR-400. The most right one in the figure is used by OSCAR 2.

Another difference to OSCAR-1 is that OSCAR-2 has pressure sensors on its feet (Figure 3.5) instead of the binary contact sensors, so a variable pressure on the robot's feet can be sensed.

Furthermore, OSCAR 2 has an accelerometer sensor used to sense the acceleration and inclination of the robot.

By experiments with OSCAR-2, National Instruments hardware [Nat06] and software was used for acquisition and pre-processing of the signals (currents from servos, feet pressure, inclination values), their graphical representation, and data logging.

Fig. 3.6 Modified HiTec HS-645 servo with wires for current and position feedback

Other important characteristic for robot OSCAR-2 is that in order to simulate a faulty situation of the robot's legs by anomaly detection experiments, there have been modifications introduced to some of the robot's legs for those particular experiments. Such experimental leg modification is shown in (Figure 3.7).

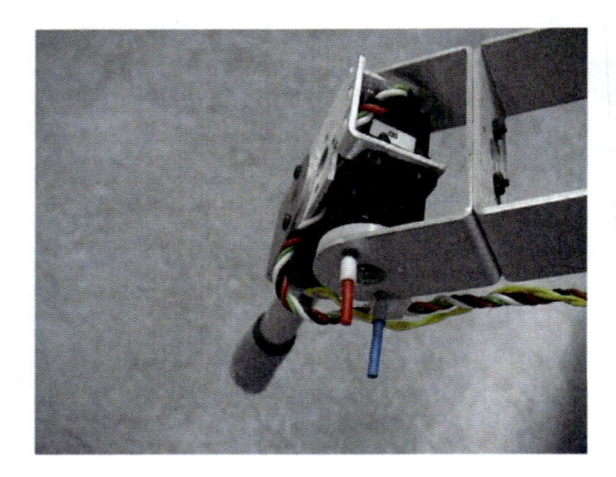

Fig. 3.7 Modification by leg of robot OSCAR-2, in order to allow simulated leg failure

The presented modification allows the robot's leg to intentionally malfunction (the inserted pins drop off) after some time of robot walking.

3.2.2.3 Hexapod Robot Demonstrator – OSCAR - 3

OSCAR-3 is similar to OSCAR-1 and OSCAR-2 in its principal construction. The difference to OSCAR-1 and OSCAR-2 is that the 18 modified HiTec HS-645 servos have additional internal electronic printed board circuits that provide servo current feedback through an I2C bus back to the computing unit. Robot OSCAR-3 doesn't have a microcontroller onboard, but instead it is connected to a PC via a USB cable. It uses the "Generic robot architecture" [Gen08] concept in order to provide a better software driver access to the robot's sensors and actuators, and therefore easier control and actuation of the robot.

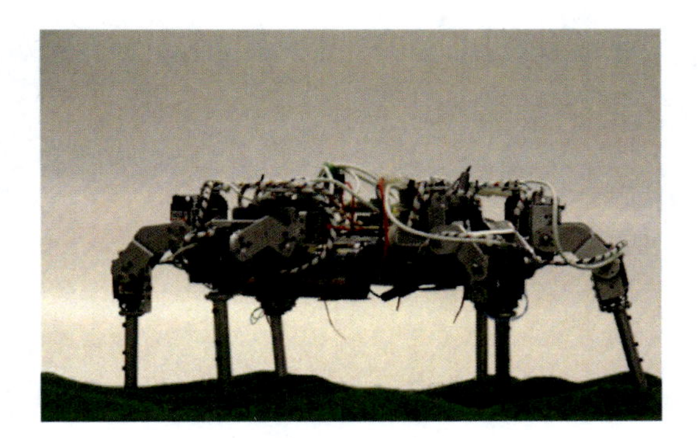

Fig. 3.8 Hexapod robot OSCAR-3.

3.2.2.4 Hexapod Robot Demonstrator – OSCAR - X

The new prototype of the OSCAR robot generation, called OSCAR-X (Figure 3.9), is built to provide a better robot research test-bed for testing the biologically inspired algorithms. In comparison to its predecessors, the OSCAR-X features a completely new design and was rebuilt from scratch.

New features of the robot include:

- Robot leg amputation mechanism: R-LEGAM [Jak09];
- Light weight glass-fiber body;
- Robot legs spatially distributed in a circle with 60 degrees between each two neighboring legs;
- Greater payload capabilities (sensors, batteries, camera, etc.) for the scientific measurements and experiments;

- Stronger digital RX-64 servos with digital feedback for their real time positions, torque levels, current levels, temperatures, etc.
- Powerful Lithium-polymer batteries for the servos and electronics;
- Weight of the body including the batteries is 7,5 kg;
- Improved foot design for better detection of the ground, complete with binary contact sensors;
- Powerful embedded system - Gumstix® "Verdex board" [GUM09] running embedded Linux;
- Usage of the "Generic robotic architecture" [Gen08] concept, to provide better software driver access to the robot's sensors and actuators and therefore easier control and actuation of the robot;
- Orientation sensor;
- Wireless camera and an additional camera servo.

Fig. 3.9 (a) Hexapod robot OSCAR-X in development stage; (b) OSCAR-X in nature; (c) Front view of robot OSCAR-X with onboard camera and additional ultrasonic sensors; (d) Top view of robot OSCAR-X.

3.2.2.4.1 Robot Leg Amputation Mechanism – R-LEGAM

The main feature of the OSCAR-X is the improved design of robot's legs, which aim for performing on-demand robot reconfiguration. Namely, the patented mechanism for robot leg amputation, R-LEGAM (DPMA-Az: 10 2009 006 934) [Jak09], is integrated for each of the OSCAR-X's legs (Figure 3.9 and Figure 3.10).

The robot's leg can be detached from the robot's body by software command. This is especially helpful when some of the legs malfunction. So instead of carrying the malfunctioned legs during the rest of the mission, the legs can be amputated to prevent any other future negative influence on the rest of the functional robotic system.

The in-situ reconfiguration of the hexapod robot OSCAR-X using biologically inspired approaches will be discussed in chapter 10.

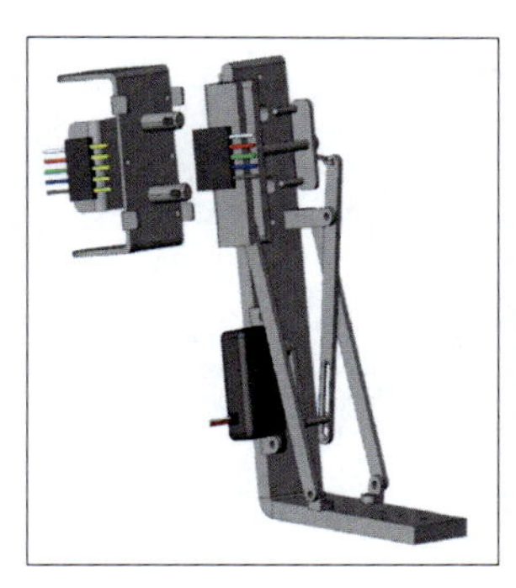
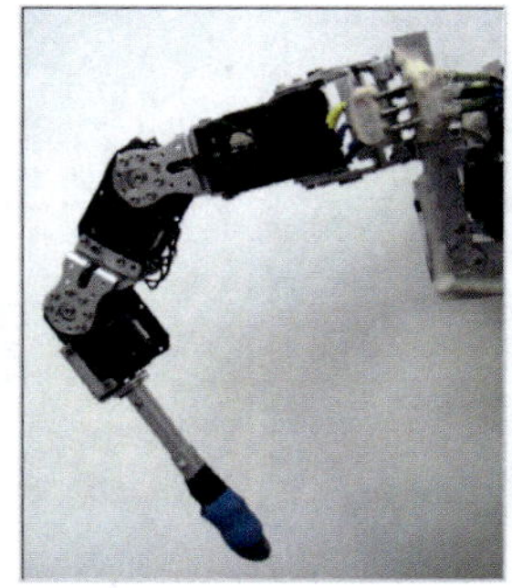
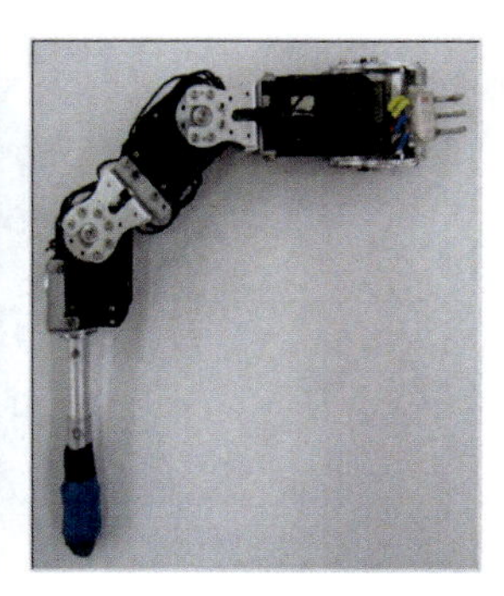

Fig. 3.10 (a) CAD design of Robot leg amputation mechanism: R-LEGAM; (b) R-LEGAM integrated on the robot's body; (c) Robot's leg detached from the robot's body using the R-LEGAM mechanism.

3.3 Humanoid Robots

Humanoid robots are robot demonstrators that have a human like appearance and are often used in robotics research, or as entertainment and service robots. Humanoid robots are therefore used to study and research the complexity of human walking and dynamic balancing, but also used for research in prosthesis development, human cognition, and human sensory information processing and perception.

They have two legs, usually two arms and a head, and are equipped with lots of actuators and sensors including accelerometers, tilt sensors, cameras, pressure sensors on their feet, ultrasonic and infra-red sensors, etc.

There are humanoid robot soccer matches organized by the RoboCup federation [Rob10], where humanoid robots autonomously play soccer. Such competitions are important for the overall research in humanoid robots and emphasize the work on developing new algorithms for humanoid robot dynamic walking and stabilization, cooperation, localization on the field, etc.

In chapter 6, research work is presented on self-stabilizing humanoid robot walking using biologically inspired algorithms.

3.4 State of the Art Humanoid Robots

There are many humanoid robots known nowadays (March 2010) that are considered state-of-the-art due to the number of features they exhibit. Here are some of the famous humanoid robots: Nasa's "Robonaut 2" (R2) [NAS10], "TOPIO 3.0" [TOS09], "ASIMO" developed by HONDA [HON07], "Albert-Hubo" [HAN05] by Hanson Robotics, and "NAO" by Aldebaran Robotics [ALD10]. (Figure 3.11)

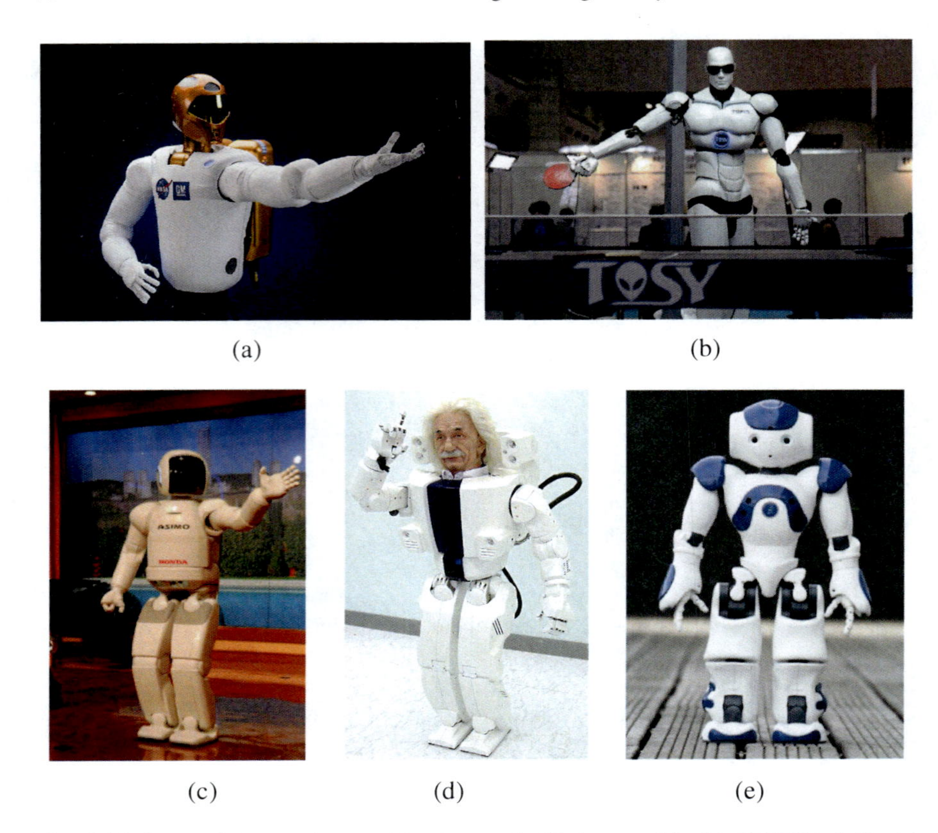

Fig. 3.11 State of the art humanoid robots: (a) "Robonaut 2"; (b) "TOPIO 3.0"; (c) "ASIMO"; (d) "Albert-Hubo"; (e) "NAO".

The robot "Robonaut2," nicknamed as "R2," is a dexterous and technologically advanced humanoid robot developed by NASA and General Motors. The goal is to have this robot accompany the space missions and work side-by-side with humans. The "R2" has a torso equipped with a head and two arms but is without legs.

"TOPIO 3.0" is the table tennis playing robot, designed and constantly improved by the company Tosio. The robot is said to use artificial intelligence algorithms to continuously improve its playing skill level.

Robot "ASIMO," developed by HONDA, is one of the most famous humanoid robots and is mostly used for entertainment purposes. This robot can detect faces, shake hands with humans, walk up and down the stairs, run, and even perform small jumps while running.

Robot "Albert-Hubo" is built on "Hubo 2," a "KHR-4" [HUB03] robot model, and is the next generation of the "KHR-3" humanoid robot with Albert Einstein's head mounted on its body. The robot is used for artificial muscle actuator research, autism therapy, cognitive science research, etc.

"NAO" is commercially available research humanoid robot platform equipped with a variety of actuators and sensors like: cameras, gyrometers, accelerometers, IR, and sonar sensors. It has a visual programming interface with which robotic movements can easily be developed. "NAO" robots are also in the RoboCup humanoid soccer games in the NAO - RoboCup standard league.

Humanoid robots come in various sizes, ranging from small robots like NAO robot - 58 cm up to full size robots like TOPIO 3.0 - 188cm size (Figure 3.11).

3.5 Humanoid Robot Demonstrator - S2-HuRo (Self Stabilizing Humanoid Robot)

The S2-HuRo robot was developed for the research on biologically inspired techniques for humanoid robot stabilized walking, presented in chapter 6. S2-HuRo is based on the "ROBONOVA-1" (Figure 3.12) [HIT06] robot kit with "HSR-8498HB" digital servos, "MRC-3024" servo control, and integrated I/O board.

Fig. 3.12 Humanoid robot "ROBONOVA-1"

This robot has been modified extensively and specially tuned, taking the final form and construction as shown in Figure 3.12.

The modified robot is called the S2-HuRo (Self Stabilizing Humanoid Robot) presented in Figure 3.13 (a)-(d). The robot is built of the following hardware elements: embedded system - Gumstix® Verdex [GUM09] represented in Figure 3.14 with wireless module and antenna, 2-axis accelerometer and gyroscope, Lithium-polymer battery pack, voltage converters for 3, 5, and 6 volts for the electronics, and binary sensor contacts (Figure 3.15). In order to decrease the weight of the robot, some servos from the "Robonova 1" arms have been removed during the robot hardware tuning, and the connection between the segments at those points is made with screws instead. The final S2-HuRo robot appears as seen in Figure 3.13.

Fig. 3.13 (a) – (d) S2-HuRo (Self Stabilizing Humanoid Robot).

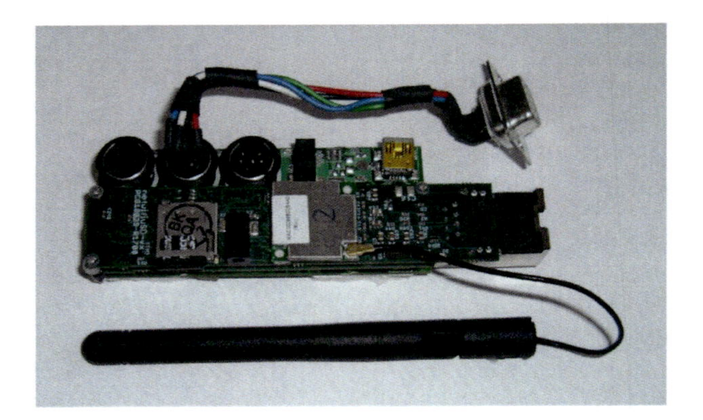

Fig. 3.14 Gumstix® Verdex embedded system with wireless LAN module, antenna, MMC card, and additional serial connector cable. View with three sensors per foot. “L” and “R” indicate the left and the right robot’s legs.

When comparing the robots shown in Figure 3.12 and Figure 3.13, differences in robot modifications can easily be spotted. By looking at the top view of the robots, the robot (Figure 3.13) (c) the two axis accelerometer sensors can be spotted under the robot's head. In the backside figures (Figure 3.13) (d) and (e) of the S2-HuRo, the case of the embedded Gumstix® Verdex system can be seen along with the wireless antenna next to the case.

On the robot's feet Lithium-polymer batteries can be spotted – two batteries per foot. The relatively light weight 5Wh LiION rechargeable batteries used by S2-HuRo for servo and electronics power supply are the same batteries used by E-pucks robots [EPU09]. The voltage from the batteries is further down-regulated by integrated voltage convertors to be compatible with the voltage operating range of the servos, Microcontroller AtMega "MRC-3024" board, and the additional electronics. The batteries are located on the robot's feet so the center of gravity of the robot is lowered and this increases the dynamic stability of the robot.

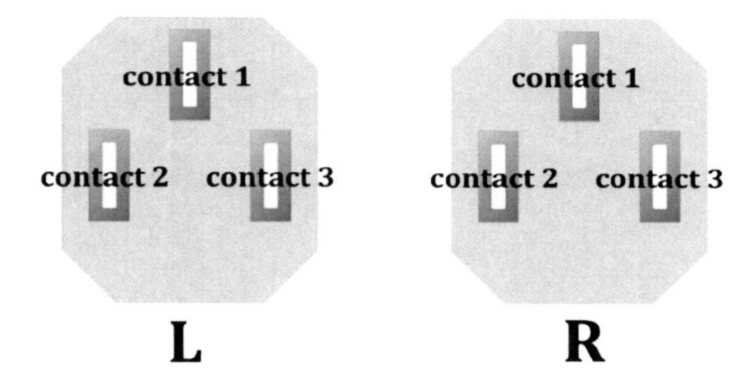

Fig. 3.15 Schematic view of binary contact sensors by the S2-HuRo feet – bottom.

Chapter 4
Biologically Inspired Robot Control Architecture

Robot control architectures are related to sensing, monitoring, and acting actions of the robots. They are an important part of each robot control and the coordination of their behaviors. There are different kinds of robotic architectures implemented by different kinds of robots.

Most of the control robot architectures are divided into following groups:

- Reactive and Subsumption / Behavior-based control architectures (schema based);
- Deliberative control architectures (hierarchical) or Sense-plan-act architectures;
- Hybrid control architectures.

These "common" types of robotic architectures are explained in the subsection that follows.

On the other hand, the research of biologically inspired robot control architectures aims at developing robot control architectures that would have life-like properties and are able to self-organize their constitutive components instead of predefining them manually. They must be able to self-optimize for their best performance and be capable of detecting the newly attached components such as sensors and actuators. They must also be able to autonomously re-configure themselves and continue with execution of their mission tasks.

Although these are all nice features to have for robot control architecture, developing such "life-like" robot control architecture is rather complicated. The tricky design of such architecture involves re-thinking aspects such as:

- The types of elements that would constitute such control architecture;
- The property of the architecture must be generic and independent of the configuration and attached hardware and electronics;
- How can an on-demand reconfiguration be achieved in such autonomic architecture so that the new components can be easily added to the running system?
- Where to place the "main control" unit or if such a unit should exist at all in such module-based architecture.

B. Jakimovski: Biologically Inspired Approaches, COSMOS 14, pp. 23–34.
springerlink.com

Questions also arise on how the decision for a particular detailed design of such architecture might influence the proper working of the other constitutive elements of the architecture. Properties such as scalability and emergence also have to be addressed.

4.1 Overview on "Standard" Types of Robot Control Architectures

In this section several common types of robot control architectures are introduced: Reactive control architecture; Subsumption control architecture; and Deliberative/reactive control architecture. Then an overview on autonomic types of architectures is given, before an explanation of the ORCA (Organic Robot Control Architecture). ORCA is on the other side directly related to the research presented in this book as well as to several ideas that are presented about improvement of robot control architecture using biologically inspired concepts.

4.1.1 Reactive and Subsumption and Behavior Based Control Architecture

By reactive control architectures (schema based), the control is stimulus-response based, where behaviors are represented by direct sensor to actuator predefined reaction mappings. (Figure 4.1) Although the speed of response by the reactive control architecture is rather high, which might be suitable for some real world scenarios where the reaction time might be very important, reactive architectures are perhaps not suitable for tasks where predictive planned outcomes should be generated.

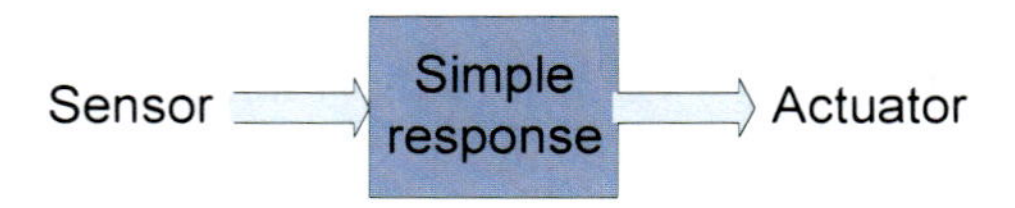

Fig. 4.1 Principle of reactive control architecture.

An alternative approach to the reactive system architecture is subsumption architecture (behavior based) introduced by Brooks in 1986 [BRO86]. This approach is based on priority behaviors organized into layers, where higher priority behaviors subsume lower priority behaviors. In subsumption architecture, the lower layer behaviors (reflexes) can inhibit higher layer behaviors. In this bottom-up fashion, the reflexes are bottom layers that can be expressed in the fastest way. The upper layers are related to higher robot control work in inhibited fashion, depending on their priority and currently executed lower actions. It is also important to mention that there is no higher level supervision in this architecture. The subsumption architecture can be useful when overall robot behavior should be dynamic, reactive, emergent. Difficulties in this architecture are related to determining the priorities of the layers constituting the architecture. An example structure of subsumption architecture is presented in Figure 4.2.

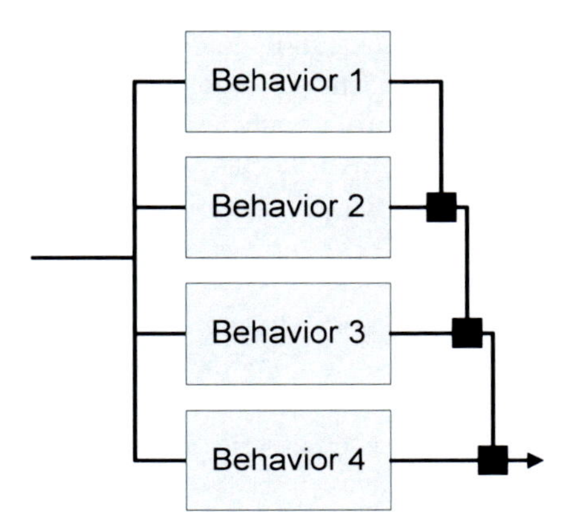

Fig. 4.2 An example structure of subsumption architecture.

4.1.2 Deliberative Control Architecture

Deliberative control architectures are based on the Sense-Plan-Act principle. For their optimal functioning, deliberative hierarchical control architectures usually need full knowledge about the environment.

By these types of control architectures, the robot first senses the environment, then plans potential solutions and considers the results when choosing appropriate actions. It is assumed that the world model is provided. The robot then executes the actions through actuators. The structure of such architecture is represented in Figure 4.3.

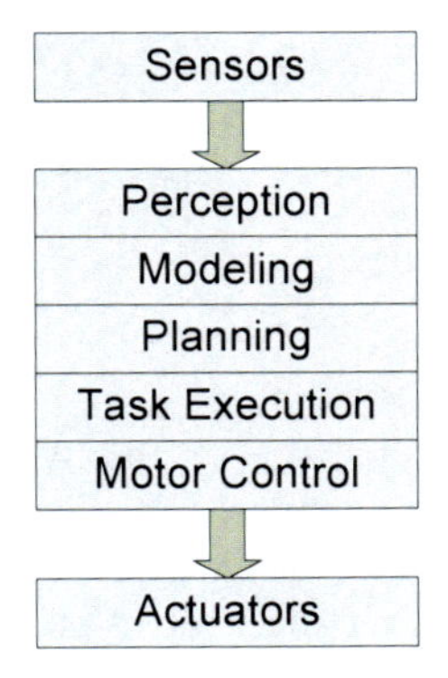

Fig. 4.3 Sense-Plan-Act model of deliberative control architecture.

The advantages of using the deliberative architecture is that in such goal oriented control architecture the goal of a given task can be achieved in a planned way. However, the difficulties or drawbacks of using such architecture are related to the re-planning phases which introduce slow response to some actions. Therefore the architecture is perhaps not suitable for tasks where fast response is needed. An additional drawback is that in case the environment changes, there also need to be changes in the control architecture, so its reaction can be compatible with the changed model of the environment.

4.1.3 Hybrid Control Architecture

One way of mitigating the limitations and drawbacks seen by the reactive and the deliberative control architectures is to combine both of the architectures into a hybrid control architecture.

A first proposal for usage of such hybrid architecture was made by Arkin [Ark87]. Since then, different kinds of hybrid robot architectures have been proposed [Con92] [LHG06]. In general, the hybrid architecture uses higher level planning in order to guide the lower level of reactive components. It is often depicted as a three layer architecture, where the top layer is the deliberative layer, operating under a slower sampling rate than the bottom layer, which is the reactive layer with a fast reaction time. The middle layer might have different interpretations and implementations for different projects, for example, aggregation of information coming from the lower layer. This control architecture can be represented as in Figure 4.4.

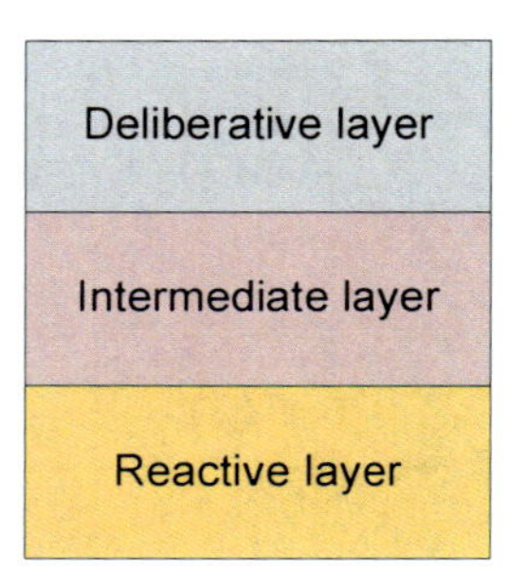

Fig. 4.4 Model of a hybrid control architecture.

The benefit of using such architecture is that it is still a goal oriented architecture where planning for the next actions occur by a deliberative layer, and at the same time "lower level" actions can be executed by the reactive layer. Therefore it can be more or less assured that the proper planned actions will be executed and also that the robot will interact better and react faster to changes in the environment.

4.2 Overview on Autonomic Control Architecture

Previously introduced robot control architectures (reactive, deliberative, and hybrid architectures) have the modules, behaviors, and tasks mostly planned, modeled, and defined in advance by a human operator. Autonomic control architectures use the idea to build architectures that will easily cope with the high complexity of the technical systems, and that dynamically adapt with respect to available resources and user needs [NaB07]. Responses taken automatically by a system without real-time human intervention are called autonomic responses [SHR06] [LVO01].

Given only high-level commands, the autonomic systems should be able to manage themselves [KeC03] in a self-governing manner. The idea of autonomic computing was first introduced by IBM in their Manifesto for Autonomic Computing [IBM01]. They proposed several features that autonomic systems should exhibit such as: self-configuration, self-healing, self-optimization, and self-protection, all of which were inspired by the human body's autonomic nervous system. These terms were explained previously in Chapter 2.4.

Autonomic systems consist of *autonomic elements* – which can build relationships with other autonomic elements and manage and influence or change their behavior in order to comply with the higher level policies defined by human operators. Such an autonomic element is represented in Figure 4.5.

As seen from the figure, each *autonomic element* has an *autonomic manager* and one or more *managed elements*. An *Autonomic manager* is associated with control of the *autonomic element* and actions like Monitoring, Analysis, Planning, Execution using a Knowledge base. *Managed elements* on the other hand represent the hardware resources like storage, the CPU, etc.

The autonomic manager controls or influences the execution of the *Managed element* and monitors its operations. Therefore it is a closed feedback loop architecture.

The concept and functionality of the autonomic control architecture differs from the "common" control architectures, and introduces the notions of self-configuration, self-healing, self-optimization, and self-protection, working towards building self-managing systems.

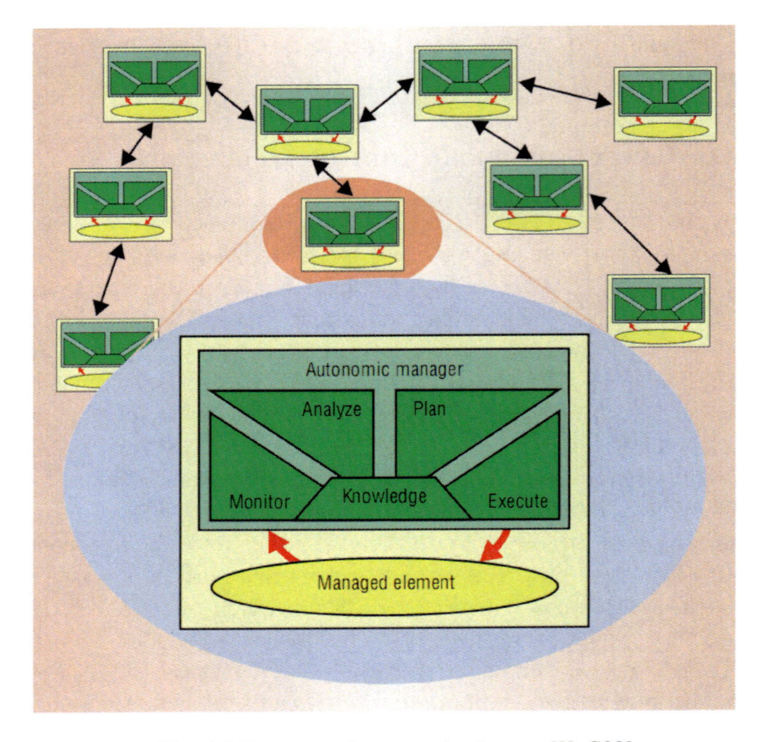

Fig. 4.5 Structure of autonomic element [KeC03].

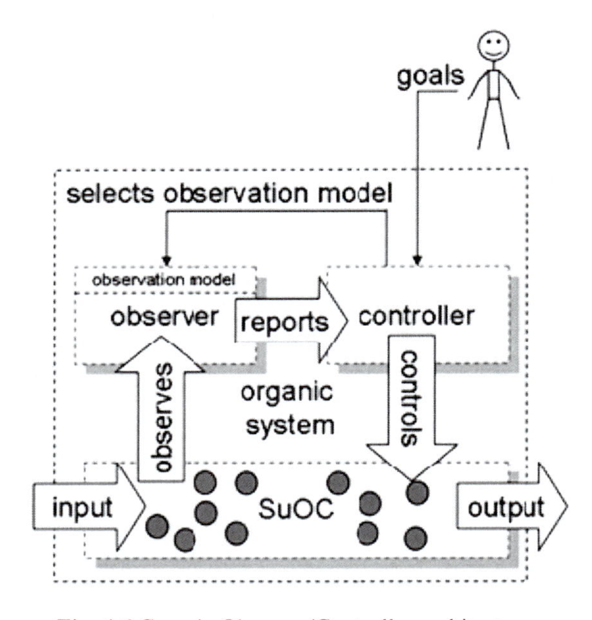

Fig. 4.6 Generic Observer/Controller architecture.

Within the Organic Computing initiative an Generic Observer/Controller architecture (Figure 4.6) has been firstly introduced in [RMB06], which incorporates Observer and Controller units, where the Observer monitors the proper behavior of Controller units and modifies their control in order to suit the predefined goals. Observer/Controller architecture is a closed feedback loop architecture.

4.3 ORCA (Organic Robot Control Architecture)

ORCA development [BMM05] is a result of Organic Computing (Chapter 2.2) research on developing hybrid robust robot control architecture that has self-x properties and at the same time provides safe and reliable functioning.

Usually it is important for control architectures that exhibit self-x properties to have the controlled emergence property. Namely, the system should be able to learn and adapt its behavior, but at the same time not demonstrate some unwanted behavior that exceeds some pre-defined constraints defined in the system's core specifications.

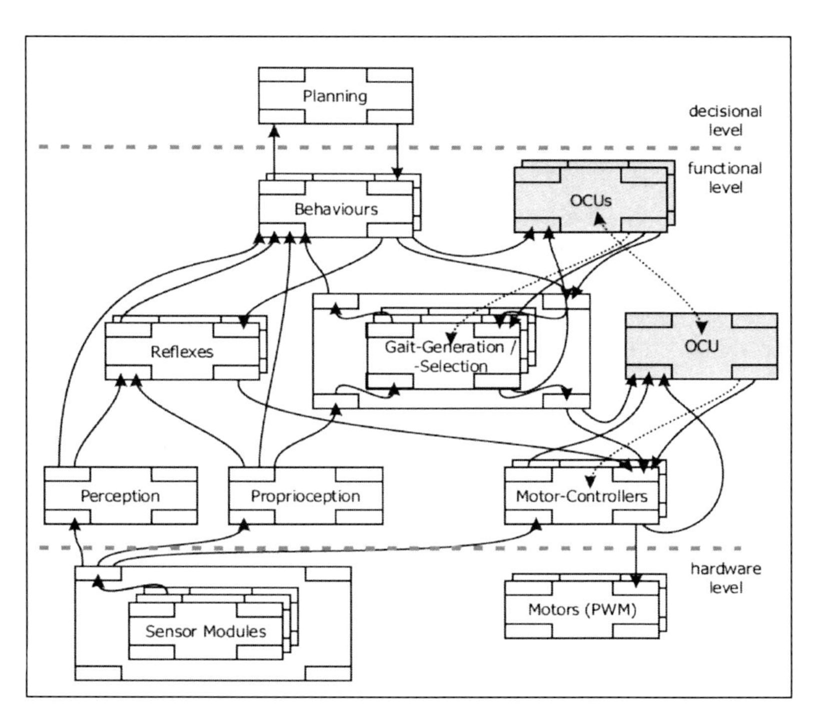

Fig. 4.7 ORCA – Organic Robot Control Architecture.

ORCA architecture is therefore built to satisfy these criteria, to provide reliable robot function & control, to have modular architecture design, and to allow for emergent properties of its constituent BCU and OCU units.

Basic Computing Units (BCUs) are basic software modules in the ORCA architecture, which may implement different functionalities related to robot control or the robot's hardware. These functionalities can be related to: sensor values acquisition, sensor data fusion, sensor information pre-processing, or for example to control a robot leg segment. *Organic Computing Units* (OCUs) are special type of units in the ORCA architecture related to monitoring tasks for the correct behaviour of BCUs and also to control them to provide counteractions in case of anomalies. ORCA architecture is presented in Figure 4.7.

ORCA is closely related to the research presented in this book, as a concept of a robot control architecture.

4.4 Distributed ORCA Architecture for Hexapod Robot Control

ORCA has also been adapted to suit the practical experiments conducted on experimental robots OSCAR-2 and OSCAR-X.

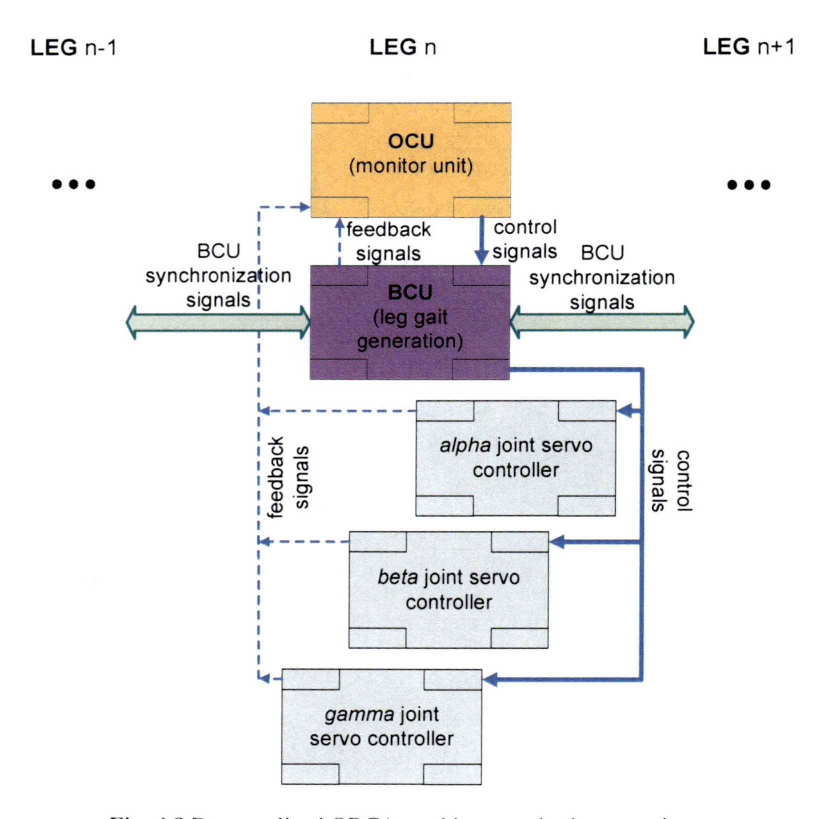

Fig. 4.8 Decentralized ORCA used in several robot experiments.

The idea behind practically implemented distributed organization of ORCA is that each leg represented in ORCA consists of a number of BCU units which can control some servo movements, and are related to some behaviors such as leg gait generation, swing/stance leg movements, etc. There can be one OCU unit that is responsible for monitoring the health status of the BCU units and providing counteractions in case of detected anomalies. There is one OCU unit per leg, and therefore several of them in the whole architecture related to monitoring the behaviors and health status of the leg units.

On the other hand, for anomaly detection purposes, a distributed ORCA can be considered with one OCU per each BCU related to servo movement. Depending on the application and task that the robot must realize, different types of distributed control architectures can be considered.

For example, the distributed ORCA architecture represented in Figure 4.8 has been used in the experiments with emergent robot walking with distributed pressure on its feet and for firefly inspired self-synchronization of walking gait of the hexapod robot.

In the following chapter information is given on exploring new ideas about ORCA architecture that would improve the robust and autonomic functioning of this control architecture.

4.5 Cell Differentiation as Biological Inspiration for Enhanced ORCA

An initial biologically inspired research was done on further enhancing the standard ORCA architecture, in order to provide means for achieving the self-x properties, like the self-organization of the modules by the robot control architecture.

A biological phenomenon called cell differentiation has been explored, more specifically the self-organization that happens in this process. This study may aid in improving the functionality of the ORCA control architecture.

First, a quick overview of the biological concept *cell differentiation* is appropriate before exploring the ideas of how these biological phenomena might be beneficial for ORCA.

4.5.1 Overview of a Biological Concept – Cell Differentiation

By biological systems, the self-organization of the components seems to be intrinsically integrated. This can be seen when observing the cells in the organs and tissues. Namely, the cells in one organism generally have some common characteristics that can be observed in all the cells in the organism. For example, they all carry the same DNA. But on the other hand, cells also do differ from each other. If we observe the brain cells, muscles cells, liver cells, etc., they all differ in their function or inner structure.

Research on types of cells called stem cells [BMT63] [SMT63] has given insight into types of cells that are able to develop into different types of cells in the body. Those stem cells are totipotent or pluripotent, and when dividing can potentially (depending on the environment) transform and become any type of cell like brain cell, muscle cell, liver cell, etc. This is done in the process of cell differentiation [STE10], in which the cell due to various environmental conditions, inter-signaling between the cells, or physical contact with neighboring cells can start to develop into a specific type of cell.

4.5.2 The Enhanced "Stem" Type ORCA Architecture

If we have a look into ORCA and its OCU and BCU units and their pre-defined functionality, we might consider these constitutive units as some sort of "already differentiated cells" in the architecture. By ORCA structure mentioned earlier, some BCU units can be related to motor control; some to control segments of the robot; others can be related to provide sensor values read from the sensors, etc. The OCUs can on the other hand be related to monitoring the individual BCU units.

The idea behind utilizing the cell differentiation concept in this context would be on introducing some "stem cell" like types of units in the architecture that can "differentiate" into appropriate module types. This would be especially useful for situations where other components like some servos and actuators (connected via bus), should be dynamically installed into the robotic system without any need of human operator intervention.

In an ordinary case such operation would require that the human operator identifies the type of module that has to be incorporated into the architecture and then programs its interface so that the component can be suitably accessed. For example, if it is an actuator, then it should be newly interfaced into the robotic system and should receive commands for actions via its interface. If it is a sensor, then it should provide data through its interface to the units that need it.

On the other hand, by the robot control architecture that has "stem" type of units, it will not have them all preprogrammed and a fixed topology of the units' interconnection. Some units at the start might be defined, for example some BCU units controlling motors in the robot. But if any additional motors are added to the robotic system, then by "fingerprint" of their functioning, they might appear to have similarities with the already defined BCU units for the motor control. In that case those "stem" type of units can be associated with BCU type interface for motor control.

Analogously this can be done also for any other extra sensors connected to the robotic system. If a newly connected sensor device has a "fingerprint" that might be similar in functionality to that of BCU units existing in the robotic system and related to sensor signal acquisition, then such extra added sensor units will get associated with BCU sensor type interface.

At first, all newly inserted hardware units are associated with some "stem" type of units in the control architecture, and then they "differentiate" within the architecture to unit types in the control architecture that best describes their functionality. When speaking of OCU units in the context of such enhanced "stem" ORCA architecture, the "stem" type of OCU units might differentiate into other different types of OCU units associated with monitoring of proper operation of "associated" BCU units.

Such “stem” type of BCU and OCU units might be interesting to explore, since upon replication and differentiation the newly generated and associated units will perhaps express new intelligent functionality of the robotic system. On the other hand these experiments with “stem” type of BCU and OCU units should be approached with great care, since there is also an open possibility that by such differentiation the OCU units might expose an undesirable property or behavior of the robotic system. There may be other techniques introduced that help to prevent this from happening, like introducing special OCU monitor units that will “guide” this process of differentiation to be compatible with the pre-defined functional requirements of the robotic system.

The enhanced “stem” ORCA architecture model is presented in Figure 4.9.

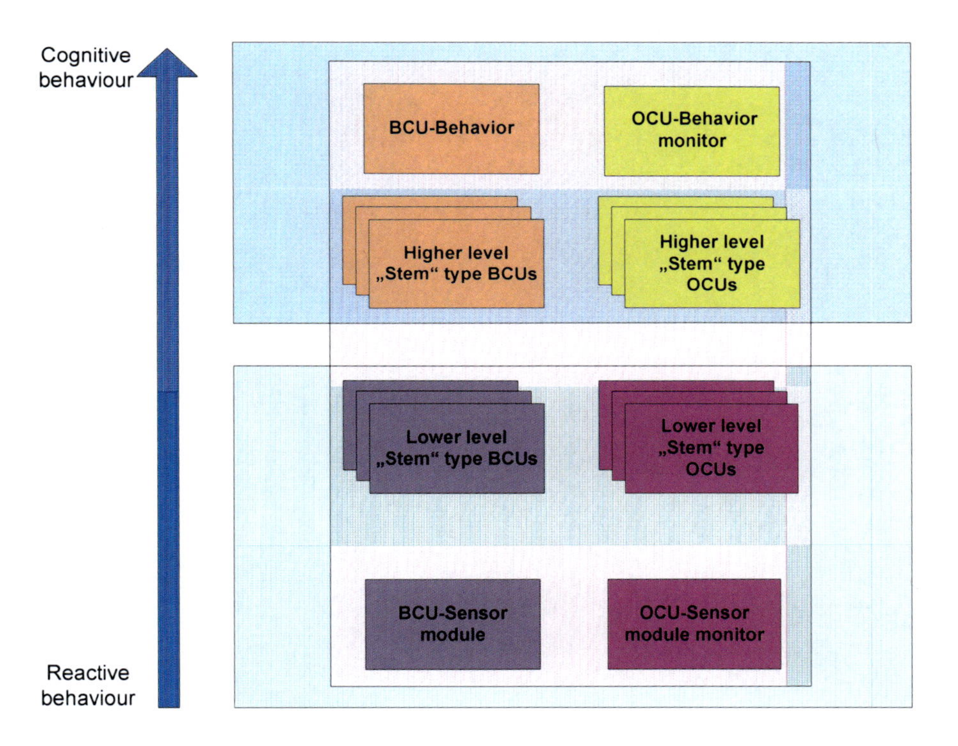

Fig. 4.9 Enhanced “stem” type ORCA.

The whole mechanism of differentiation of the BCU and OCU units should be explored, in the sense of defining a method for converting the “stem” type BCU or OCU units. For example, they may be influenced by the neighboring units to become a completely defined unit within the control architecture with a specific function and interface.

One idea would be that the data oriented sequences originating from the OCU and BCU units can be compared to the data streams originating from new "stem" units. This might help the "stem" OCU or BCU units to be influenced to change their type to that kind of BCU or OCU units that have similar data streams. This would be analogue to the biological counterpart of cell differentiation, where the different types of cells have or release slightly different types of molecules near their vicinity, which differ from the molecules released in their surroundings by other type of cells in the organism.

By the enhanced "stem" type ORCA organized in a bottom-up fashion, the differentiation that occurs by the "stem" type of lower level BCU units and lower level OCU monitoring units, is related more to the reactive behavior of the system. In contrast, the higher level differentiated BCU and the higher level OCU units (provided for their monitoring) are related to expressing the cognitive behavior of the system. The rectangles show an example on how such elements would communicate with each other in such enhanced "stem" type ORCA.

Chapter 5
Biologically Inspired Approaches for Locomotion of a Hexapod Robot OSCAR

Different types of walking gait generation approaches have been considered for walking by various multi-legged robots. Some of them are based on mathematical formulations [PaH09] or inverse kinematic models [ShT07] trying to mathematically model and describe the kinematics of the robot movements and also the interaction of the robot with the environment. This may prove difficult since completely modeling the robot and its interaction with the environment and other environmental influences on proper working of robotic components is very complex.

Another kind of approaches are biologically inspired. The biologically inspired approach CPG (Central Pattern Generator) is based on walking patterns generated by neural-networks [Mat87] seen by animals and insects. This is acquired for producing walking movement by multi-legged walking robots [ITG09].

Another type of biologically inspired approaches can exhibit self-x properties such as: self-organization or self-reconfiguration, similar to self-organization properties seen in biological systems.

Before describing in detail the research done about such self-organizing walking gait patterns based on emergence, several characteristics are given for locomotion by insects and CPG based types of "common" walking gaits seen by animals and insects.

5.1 Characteristics of Locomotion Seen by Insects and Animals - Applied to Robotics Domain

Observations and research done on insects, arthropods and animals has provided new insights on the locomotion seen in nature and on how living organisms generate their movement patterns. These observations have been useful for developing basic insect movements.

For example, research on the stick insect (Carausius morosus) [Cru76] walking has given new information about the functionality of the leg segments and their relation to protraction and retraction; elevation and depression; extension and flexion.

B. Jakimovski: Biologically Inspired Approaches, COSMOS 14, pp. 35–66.
springerlink.com

There are two phases that are characteristic and essential for leg movement by the insect: swing and stance phases. There are also two remarkable positions for these movements – the posterior extreme position (PEP) – the position of the leg on the ground when the leg is at the end of the power stroke; and the anterior extreme position (AEP) – the position of the leg on the ground when the leg is at the end of its return stroke.

In the swing phase, the leg is moving from the PEP to AEP, shown in Figure 5.1.1 – movement which precedes the movement of the leg over the ground. In the stance phase, the leg is moving from AEP to PEP, which produces the thrust that moves the insect over the ground. The swing and stance phases are characterized with the length of their respective trajectories. The lengths of swing and stance trajectories have direct influence on the speed with which the insect is moving over the terrain. The longer the trajectories of swing and stance, the bigger the distance travelled by the leg on the ground, and vice-versa, the shorter the trajectories, the shorter the distance travelled on the ground. The length of swing and stance phases by one leg in combination with swing and stance phases of other legs may influence the insect to turn around its vertical axis.

As represented with circles in Figure 5.1, the leg of the insect can be considered to be a 3 degrees of freedom (3 DOF) system. Namely, the leg consists of three segments connected via joints, providing basic movements for protraction and retraction; elevation and depression; extension and flexion.

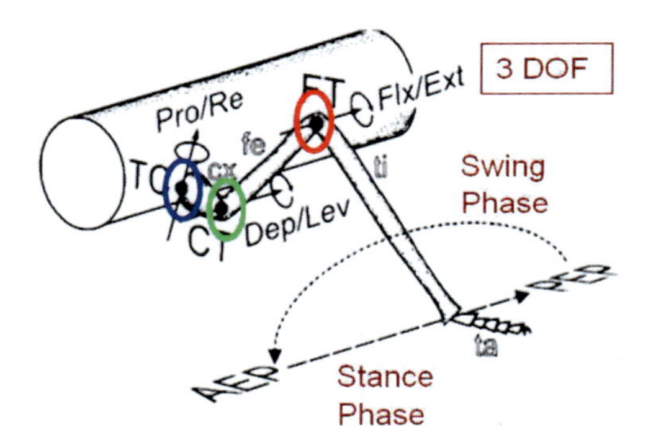

Fig. 5.1 Swing and stance phases of an insect's leg.

The same principle seen by leg structure and movement by insects has been used for constructing and providing locomotion of the legs of multi-legged walking robots. In Figure 5.2 (a), a 3 DOF structure is indicated by one of the robot's legs, which is similar to structure seen by the insect's leg – the circles represent the joints and the degrees of freedom. Servo "Alpha" is the nearest to the body on each of the legs; servo "Beta" is next to it, down the leg; servo "Gamma" is the last servo on the leg outward from the body. The swing and stance phases and their trajectories by robot's leg are presented in Figure 5.2 (b), which are similar to swing and stance phases seen by an insect's leg (Figure 5.1).

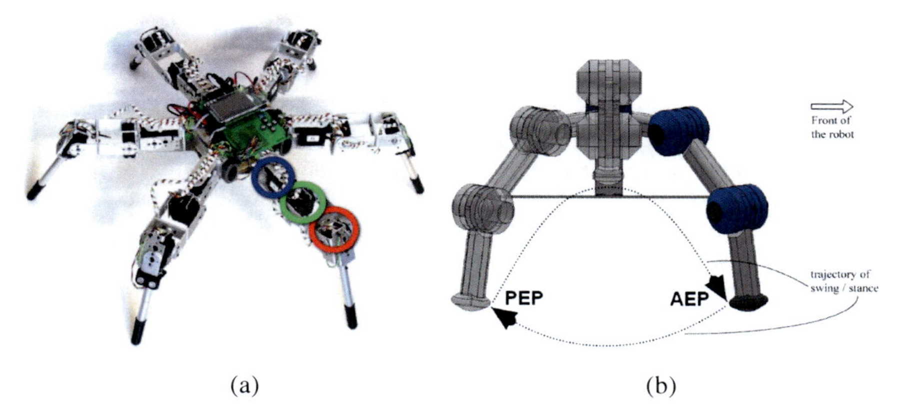

Fig. 5.2 (a) 3 DOF structure represented with circles on one of the robot's legs; (b) Swing and stance phases and their trajectories of the robot's leg.

5.2 Central Pattern Generators (CPG)

Research on walking gait patterns has shown that neural networks called Central Pattern Generators or CPG are located in the neural systems below the brain stem [Gri81] in the spinal cord and are responsible for generating and modulating the walking patterns and others specific to rhythmic motions.

Mathematical models for CPG are proposed in [Mat87] where the CPGs consisted of two neurons, where each neuron receives an excitatory tonic input, a mutual inhibition, and an external inhibitory input. At the end the output is

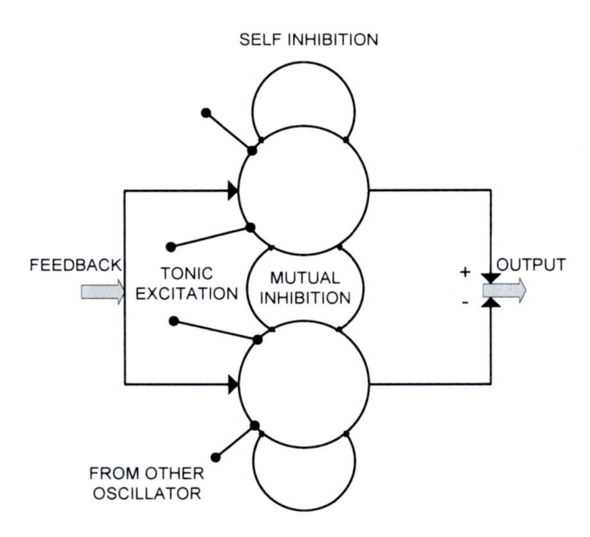

Fig. 5.3 CPG model with two inhibiting neurons.

generated as a joint interaction of such inhibiting neurons. The mutual self-inhibition is useful to induce a stable relaxation without any external input, therefore an oscillation can be provided. Figure 5.3 represents the schematic of such a model of two inhibiting neurons.

The neurons which mutually inhibit themselves can be related to flexion or extension movements of the legs, which are important for walking. It has been also shown that activities of CPGs can be modified by sensory feedback [CoB99] and reflexes, which gives insight on how the rhythmic and reflex movements can be coupled. Also Holk Cruse together with Friderich Pfeiffer did pioneer work on a better understanding of biological control systems for 6 legged walking machines (as well as Büschges, Berns, Ilg, Albiez,..).

5.2.1 Common Observed Gaits by Insects

Several walking gaits, which might be generated by an insect's CPGs have been observed: Wave gait, Ripple gait, and Tripod gait.

- *Wave gait* – is the slowest gait where one leg is in swing phase (in the air) while the other legs are in the stance phase (on the ground). This gait is characterized as the most stable one, since all the other legs are on the ground and supporting the robot's body. However, this is also the slowest walking gait, since only one leg is in the air at a time, while the others are on the ground.
- In *Ripple gait* – there are two independent gaits from both sides of the body. The stance phase is usually double the swing phase and the opposite legs are 180 degrees out of phase.
- The research on walking gaits of six legged insects, for example a cockroach [SpM79], has indicated that the *Tripod* walking gait provides the six-legged insect with the fastest speed over the ground. Tripod walking gait is a gait by which at any moment of time three of the robot's legs are in the swing phase, while the other three legs are in the stance phase. The three legs on the ground provide the insect with static and dynamic stability while walking.

These gaits are represented in Figure 5.4 (a), (b), (c), going from the slowest one – wave gait, then ripple gait, and lastly the fastest: tripod gait.

The black filled bars indicate the stance phase, while the non-filled ones represent the legs during swing phases. On the symbolic represented insect in Figure 5.4, the symbols L1, L2, L3 indicate the legs on the left side of the body with their respective numbering and position on the body. The symbols R1, R2, R3 indicate the legs on the right side of the body with their respective numbering and position on the body. The symbols for left and right legs and their numberings are located on the vertical axis, so it can be seen for which legs what type of leg movement is taking place. The horizontal axis represents the time domain.

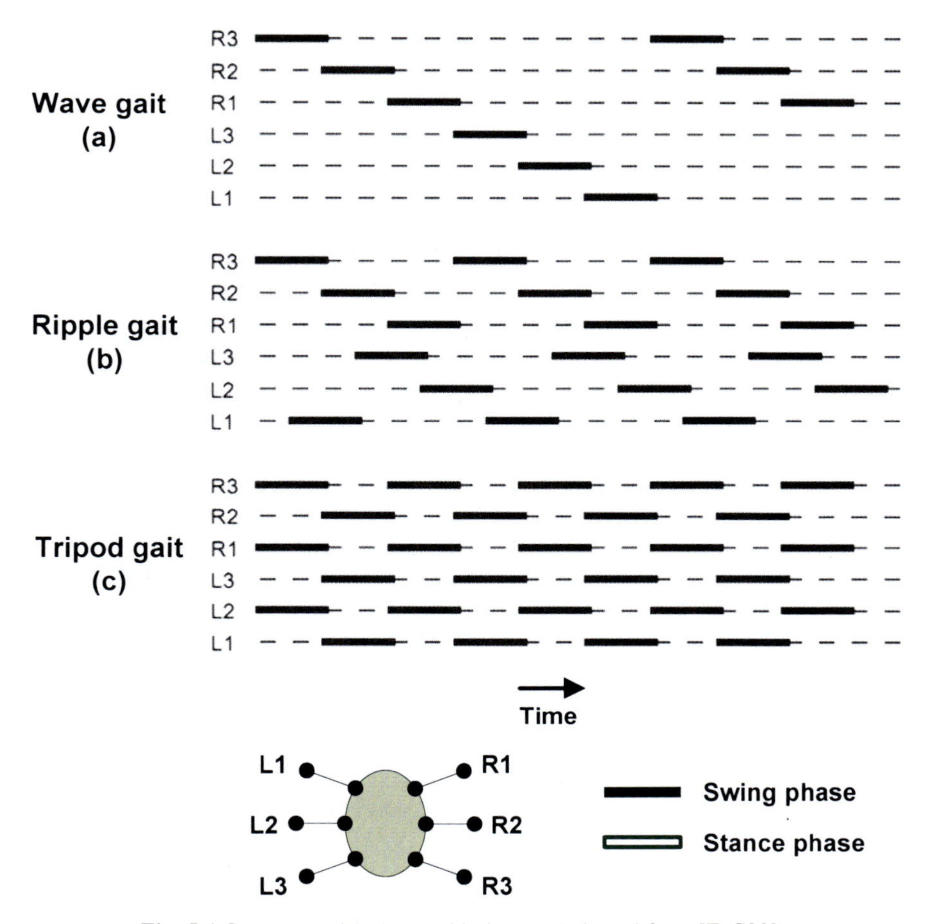

Fig. 5.4 Common gaits observed in insects (adapted from [FaC93]).

The presented insect walking gaits have been successfully applied for generating robot walking for the domain of multi-leg walking robots patterns [IYS06] [YAL06] [MDB07].

The fastest walking gait, tripod gait, has often been applied to robotic projects to generate walking motion by the six legged walking robots. The tripod gait has also been used within this research to provide the walking for the robot OSCAR-X in some of the experiments which will be explained later in chapter 5.4.

5.3 Experiments with Self-organizing Emergent Robot Walking Gait with Distributed Pressure on Robot's Feet

In the previous chapters 2.4 and 2.5, the terms about self-x properties and the emergence were explained as being important for exploration and implementation in robotic systems. Namely, the robots that exhibit self-x properties would be able to adjust their walking gaits in respect to the changing environment in an autonomic manner. By implementing self-organizing walking gaits, the robots could be also more stable during their walking and distribute their weight uniformly over their legs. Self-organizing robot walking gait patterns mean also that the walking gaits of the robot will not be pre-programmed, which minimizes the human effort to develop a functional walking robot. This also enables another property that cannot be fully pre-programmed in advance - the walking gait can adapt to the environmental circumstances.

For example in [AWY99], the phase of a CPG oscillator dedicated to each leg is controlled by calculating the energy consumption of the leg's joints. Therefore gait pattern changes adaptively and in an emergent way with the robot's walking velocity. By research presented in [TTK01], the gait pattern emerges through modulation of phases of the oscillators with mutual interactions and the feedback signals sent from the touch sensors on the feet of the robot.
Additional experimentation with emergent walking has been done and described in [EML06], where the coordination of legs has been done with the following rule: a leg is only allowed to swing when its neighboring legs perceive a ground contact. In this research, the ground contact was perceived with a binary sensor. With previously pre-coding the swing and stance phases by each of the robot's leg and implementing this rule, the hexapod robot performs emergent walking. This research has been conducted on the robot platform OSCAR-1 (section 3.2.2.1).

Additional research, extending the last research, was made on developing self-organizing, emergent walking gait for the hexapod robot that would enable the robot to have stable walking and at the same time distributed pressure on its legs – meaning that the weight of the robot is distributed evenly on the robot's legs by its walking. This aims to having the robot walk over a terrain with more stability. The research was done on OSCAR-2 (section 3.2.2.2) with pressure sensors on the robot's feet. The advantage of having pressure sensors in comparison of binary sensors on the robot's feet is that the robot could sense the weight or pressure being present on its legs. Which changes dynamically during its walking.

The rule for generating emergent walking with pressure distribution is that the leg is allowed to swing only if the neighboring legs have achieved "more or less" equal pressure on their feet when entering their stance phase. Another difference to the previous research is that the swing phases are pre-programmed, but the stance phases are not hard coded. This means the leg pushes its lower leg segment and foot towards the ground only until "enough good" pressure is reached in comparison with its neighboring legs. The threshold value for measuring if the feet have similar pressure has been experimentally determined (various pressure sensors have different sensitivity levels over the measurement range).

For practical demonstrations, the self-organized emergent robot walking was also implemented into the ORCA architecture (explained in chapter 4.3) by the hexapod robot OSCAR-X, and more precisely within the distributed ORCA architecture (explained in chapter 4.4).

In such control architecture on the robot's level there is:

- One OCU (Organic Control Unit) – related to monitoring of the proper function of BCU units.

The robot's architecture has decentralized organization where each of the six legs of the robot's control architecture consists of:

- One BCU related to leg gait generation, which also communicates with other BCUs located at the same "level" in other neighbouring legs and are also related to leg gait generation;
- Three BCUs representing the three servo segments in each leg.

In such decentralized control architecture, each BCU is related to gait generation and in combination with other BCU units of other legs they all aid in generating a self-organized emergent robot walking gait.

To better describe when such self-organizing gait with pressure distribution emerges, two common patterns have been observed by a leg and its leg neighbors.

One example case where such distribution of pressure by the legs occurs is represented in Figure 5.5. Only three neighboring robot legs are shown to better describe the distribution of pressure that happens by using such a rule. Such a pressure distribution situation can be found in any group of three neighboring legs of the robot. The situation in Figure 5.5 shows one leg (Leg2) is on the ground in its stance phase and the other two neighboring legs (Leg1, Leg3) are ending their swing phases and entering their stance phases.

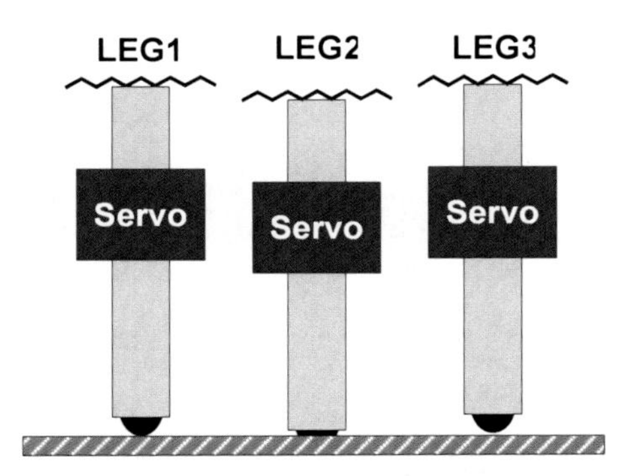

Fig. 5.5 Example case – 1: Start of distribution of pressure on the feet by the robot's legs.

To better describe the situation, Figure 5.3.2 shows the pressure on the three feet on the robot's legs, indicated with gray tones in rectangles. The darker the region, the more pressure is on the foot of the leg. Three moments are observed by distribution of pressure on the robot's feet – represented in Figure 5.6 (a, b, c).

- If one leg (Leg2) is on the ground in its stance phase and the other two neighboring legs (Leg1, Leg3) are ending their swing phases and entering their stance phases. (Figure 5.6 (a));

- One leg (Leg2) is between the legs (Leg1, Leg3) that both enter in stance phase. This allows Leg2 to swing only if both of the two neighboring reach "more or less" the same foot pressure as the neighboring middle leg in the stance phase during the start of their stance phase,. (Figure 5.6 (b));
- The legs (Leg1, Leg2, Leg3) have "more or less" equal distributed pressure on their feet (Figure 5.6 (c)), meaning that the leg (Leg2) between them can start its swing phase.

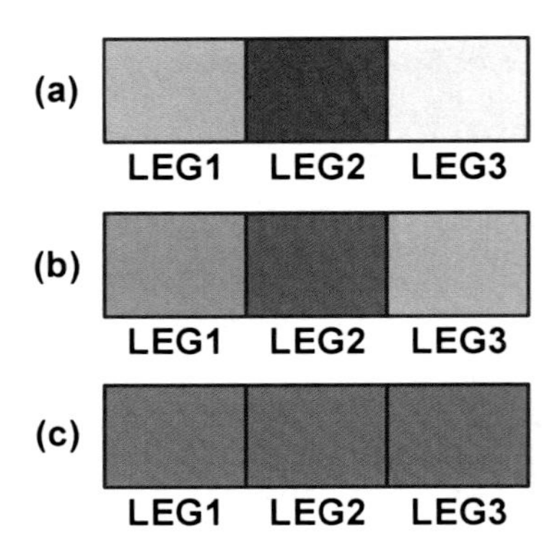

Fig. 5.6 Visualization of case when the two legs Leg1, Leg3 are entering from swing into the stance phase, while leg Leg2 in its stance phase supports them till "good enough" pressure is reached by both of the neighboring legs. Gray tones indicate the pressure on the feet of the robot's leg. Darker gray colors indicate higher pressure on the robot's foot.

Another example case is when such distribution of pressure on the legs happens, represented in Figure 5.7. Only three legs are shown to describe the distribution of pressure that happens by using such a rule. Such pressure distribution can be found in any group of three neighboring legs of the robot.

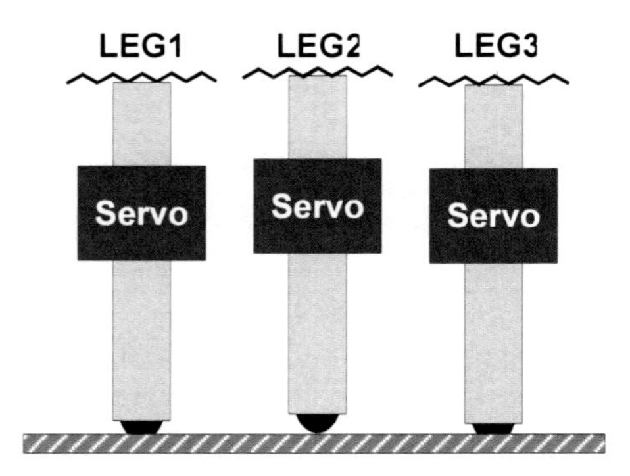

Fig. 5.7 Example case – 2: Start of distribution of pressure on the feet by the robot's legs.

Situation in Figure 5.7 shows two legs (Leg1, Leg3) that are on the ground and finishing with their stance phases and their common neighboring leg (Leg2) is ending its swing phase and enters its stance phase.

To better describe the second case, Figure 5.8 shows the pressure distribution on the feet of the robot's legs, indicated with gray tones in rectangles. The darker the region, the more pressure is on the foot on each leg. Three moments are again observed by distribution of the pressure on the robot's feet – represented in Figure 5.8 (a, b, c).

- If two legs (Leg1, Leg3) are on the ground and finishing with their *stance* phases and their common neighboring leg (Leg2) is ending its *swing* phase and entering its *stance* phase. (Figure 5.8 (a));
- The leg (Leg2) is between both legs (Leg1, Leg3) that finish their *stance* phases. The legs (Leg1, Leg3) are allowed to *swing* only if the common neighboring leg (Leg2) during the start of its stance phase reaches "more or less" the same feet pressure like the neighboring legs in their stance phases. (Figure 5.8 (b));
- The legs (Leg1, Leg2, Leg3) have "more or less" equally distributed pressure on their feet (Figure 5.8 (c)), meaning that the legs (Leg1, Leg3) can start with their *swing* phases.

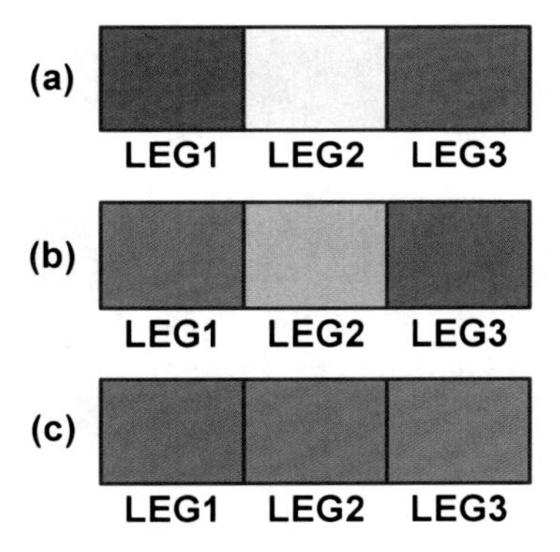

Fig. 5.8 Visualization of case when the two legs Leg1, Leg3 are ending with their stance phases, while leg Leg2 is finishing with its swing phase and entering its stance phase. The legs Leg1, Leg3 supports the Leg2 till "good enough" pressure is reached so the pressure on their feet gets more or less equally distributed. Gray tones indicate the pressure on the feet of the robot's leg. Darker gray colors indicate higher pressure on the robot's foot.

Here are some observations done during the real case scenario experiment of the robot's walking with distributed pressure by emergent walking.

In Figure 5.9., the robot's gait and walking patterns are also influenced by performing the distribution of pressure on its feet. This is depicted with:

- 6 legs on the ground; (short transition between swing and stance phases); Figure 5.9 - (a)
- 1 leg in the air, the other 5 on the ground; Figure 5.9 - (b)
- 2 legs in the air, the other 4 on the ground; Figure 5.9 - (c)
- 3 legs in the air, the other 3 on the ground. Figure 5.9 - (d)

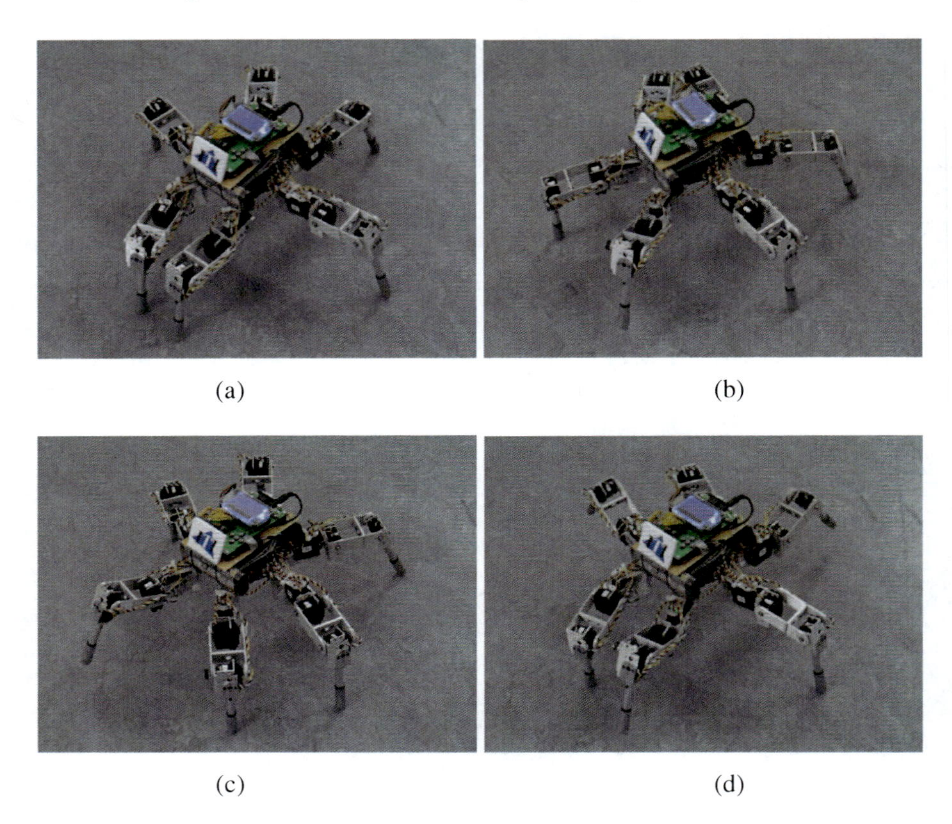

(a) (b)

(c) (d)

Fig. 5.9 Observation of patterns in emergent walking with dynamically distributed pressure on the robot's legs: (a) 6 legs on the ground; (b) 1 leg in the air, the other 5 on the ground; (c) 2 legs in the air, the other 4 on the ground; (d) 3 legs in the air, the other 3 on the ground.

From the experiments done with robots walking with distributed pressure on its feet, it can be concluded that the gait, swing, and stance phases of the robot's legs change during the robot's walking in a rather unpredictable way. At one point in time, one leg can be in its swing phase while the other legs are in their stance phases, and at some other point in time two legs are in the air while the other four are on the ground. Sometimes the gait also changes to tripod gait, when three legs are in the air and the other three legs are on the ground.

It can be concluded that the robot using this emergent kind of walking gait is walking stably over the ground, without tipping over and with a gait that can change on demand from the slowest walking gait to the fastest walking gait (tripod gait).

The positive side of using this walking approach is that the walking can be implemented in a rather simple way by the multi-legged robots using only one simple rule. The approach is also suitable for implementation on 4 or more legged robots, and is independent of the number of legs the robot has (so long the number is greater than 4).

One thing that can be improved in the future by this approach is that the algorithm for walking can be enhanced with an algorithm for keeping the heading of the robot. This would be useful for keep the robot walking in the proposed direction. Such an algorithm would have to make synergy of the rule that enables emergent walking with distribute pressure on the robot's feet. Perhaps there would also be some other new rules that would prevent some legs from lifting during particular moments of walking so the robot would avoid turning off the track. This algorithm would also have to have feedback data from the compass, inertial navigation system, or GPS sensor so the navigation of the robot can be done properly.

5.4 Firefly Inspired Synchronization of a Robot's Walking Gait

In this chapter, investigations are explained concerning applying biologically inspired synchronization for achieving self-synchronization for gait pattern parameters of hexapod robot.

The idea behind the synchronization of the gait patterns is that sometimes the walking gait pattern of the robot can change while the robot is walking. For example, this may happen when the emergent walking pattern is used by the robot's walking and a situations occurs where the leg is approaching a rock and a sensor (ultrasonic, infrared, etc.) related to that leg's movement "informs" the leg that it should alter the length of its stance/swing phase to overcome the obstacle. In that case the other robot's legs must also alter their stance/swing phases to remain synchronized. Perhaps a similar situation may occur when the servo related to protraction of the leg gets somehow blocked and can only move within a very limited range. In that case the other legs of the robot must also synchronize the length of their stance/swing phases.

One way of introducing change to walking gait patterns and their synchrony is by using Central Pattern Generators (CPGs) (CPGs were introduced in 5.2). Namely by modification of their mutual inhibition and use of sensory feedback, the walking gait and synchrony of the robot's legs gets modified. However, developing CPGs often entails cumbersome modeling and the CPG developed is also specifically made to suit the robot's design, hardware, etc., and perhaps not easily transferable to other walking robots.

The new synchronization approach researched and elaborated here is therefore related to introducing a self-synchronization approach which avoids cumbersome modeling and that can be easily transferred to other kinds of multi-legged robots, such as four-legged, eight-legged and the like.

Before introducing the self-synchronization approach which is based on firefly flashing synchronization, a short introduction is given about the firefly flashing and firefly coupled oscillator principle seen in nature.

5.4.1 Firefly Coupled Oscillators Principle

Synchrony as one type of emergence is defined as collective organized behavior that occurs in populations of coupled oscillators [ZHH98]. Synchrony can also be observed in nature by fireflies and their flashing [Buc88] [Bot95]. (Figure 5.10)

Neuropsychological studies of the mechanism of flashing within fireflies [CaS78] [BuM67] has shown that rhythmic flashing of the male fireflies is controlled by a neural timing mechanism in the brain that gives a constant frequency of the flashing. Several studies [Buc37] [Mag67] have shown that external light, such as the light from other neighboring fireflies has an effect on firefly's flashing rhythm. In references [HCB71] [Win67] [BuB76], results from experiments were presented that suggest that the external flash signal received from another firefly resets the flash-timing oscillator in the brain and therefore provides a mechanism for synchronization of fireflies flashing.

Fig. 5.10 Fireflies flashing.

In general, when one firefly sees the flashing of another neighboring firefly it shifts its rhythm of flashing in order to get in synchrony with that firefly's flashing. As a result, a synchrony in firefly flashing takes place.

The shifts of flashing rhythms by the fireflies are associated with late and early resets of the firefly flashing – as presented in Figure 5.11.

The biologically inspired firefly coupled oscillator principle has been practically applied to problems in various computer science domains, mostly related to synchronization in computer networks [BBJ07] [NAS08] [YuT08].

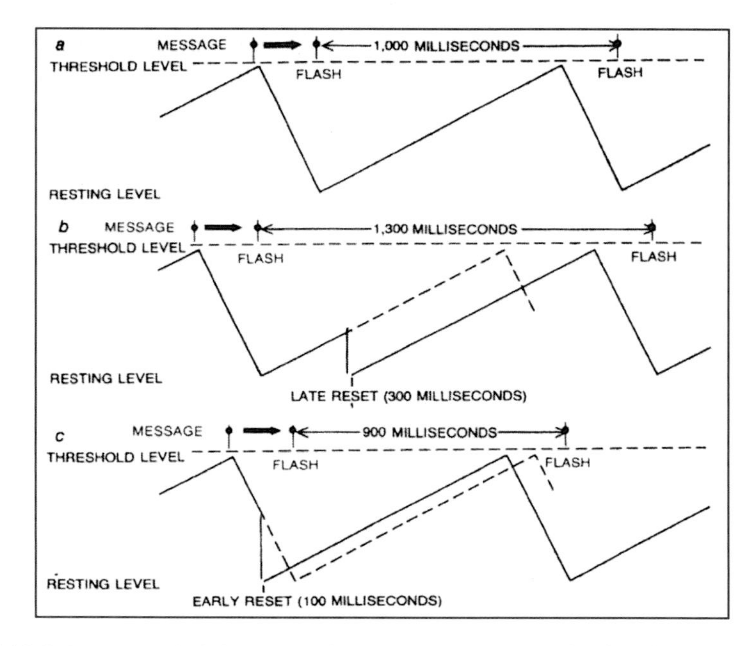

Fig. 5.11 Pulse coupled biological oscillator concept seen by fireflies in nature and the shifts of their rhythm. [CDF01].

For the robotics domain, in the field of multi-robotic systems [WHK06] research has been done on synchronizing the behavior of multiple robots. A firefly inspired walking robot gait pattern synchronization by a walking hexapod robotic system is presented in this chapter.

5.4.2 Concept for Robot Walking Gait Self-synchronization by Using Firefly Synchronization

By transferring the self-synchronization firefly concept to the multi-legged robot walking gait synchronization, there were some adaptations that had to be considered. In the research done, the robot's legs have been considered as individual units that can interact like fireflies and synchronize their gait patterns.

The concept developed for firefly self-synchronization applied to robot walking gait synchronization can be generally described with two rules:

- When one firefly sees another neighbor firefly flashing it shifts its rhythm of flashing in order to get in synchrony with the other firefly's flashing, which on a hexapod robot would be translated that when one leg changes (by shortening or prolonging) its own gait, then the other neighboring legs also adapt to this change by shortening or prolonging of their own gait.
- Prolongation takes place only in the stance phase of the leg, and the shortening of the gait takes place only in the swing phase of the leg.

As a result of applying these rules, the walking gait length of each of the robot's legs adapts to the change of walking gait of the neighboring legs. If we take into account the emergent effect, the synchronization of walking gait patterns of the robot is achieved without using global coordination for the change of the gait pattern for each of the robot's legs.

The following is a description of this concept, which is also represented in Figure 5.12.

For better understanding, a model of a hexapod robot with its legs numbered and the length of their swing/stance phases is shown in Figure 5.13.

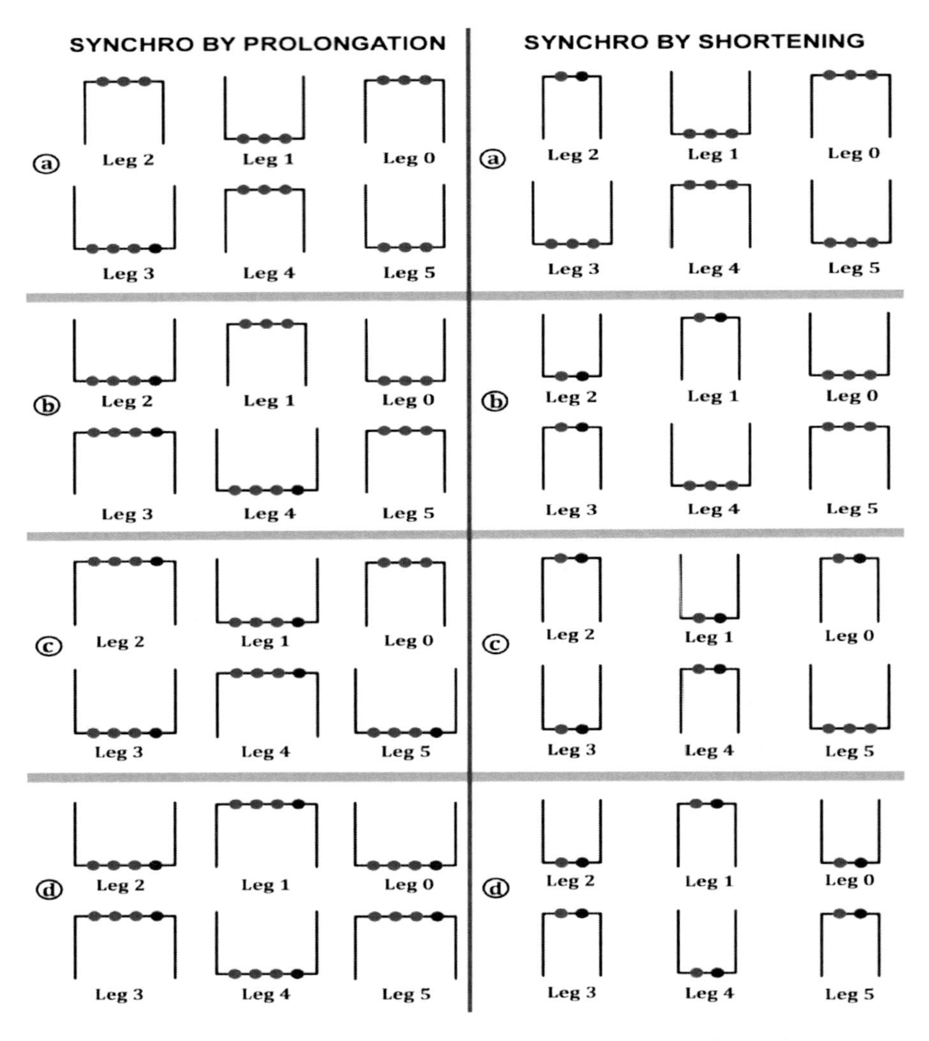

Fig. 5.12 Self-synchronization by shortening and prolongation of walking gait patterns.

The concept described here and in [JMM10] explains the self-synchronization of gait pattern lengths by a hexapod robot, although it can also be applied to other kinds of multi legged robots: four legged, eight legged, etc.

As can be seen in Figure 5.12, there are two columns: "Synchro By Prolongation" and "Synchro By Shortening" describing the concept of self-synchronization when the length of the stance phase is prolonged or when the length of the swing phase is shortened. Each of the columns has a, b, c, and d figure subparts showing how the synchronization of the gait patterns takes place by each of the legs. The lines at the legs show if the leg is in a stance phase with a "U" shaped line, or in a swing phase a "∩" shape. The number of dots on the line represents the duration of the stance or swing phases. For example 3 dots will represent 3 length units of the swing and stance phase. The length units of the real robot movement can be translated into movement degrees of the leg. For example a length of 3 units may represent 30 degrees of protraction leg movement, and 4 units may represent 40 degrees.

For describing this concept, it is assumed that all the legs at the start have the same swing and stance parameters, i.e. 3 length units. In Figure 5.12 column "Synchro By Prolongation" subfigure (a) shows the leg number 3 prolongs its stance phase from 3 length units to 4 length units. In subfigure (b) in the same column, using the firefly inspired synchronization the neighbor legs numbered 2 and 4 in their stance phase synchronize their gait length from 3 length units to the length of the gait of leg 3 i.e. 4 length units. In subfigure (c) in the same column, the synchronization wave spreads further to the other neighboring legs numbered 1 and 5 which also synchronize their stance phase length from 3 length units to 4 length units.

In subfigure (d), the leg numbered 0 in its stance phase synchronizes its length parameter from 3 length units to 4 length units. With this step, the self-synchronization of the walking gait pattern is finished and the robot continues to walk further with all legs having a gait of 4 length units. This results in a greater speed of the robot over the ground.

In Figure 5.12 in the column "Synchro By Shortening," a self-synchronization of the gait pattern is shown where the legs are shortening the length of the gait within their swing phases. In subfigure (a) in that column, the leg number 2 decreases its gait length in the swing phase from 3 length units to 2 length units. In the subfigure (b) in the same column, the next cycle is shown where the other neighbor legs numbered 1 and 3 decrease their gait lengths during their swing phases from 3 length units to 2 length units. In subfigure (c), the legs numbered 0 and 4 are the next ones that shorten their gait length from 3 length units to 2 length units. At the end leg 5 synchronizes and shortens its gait length from 3 length units to 2 length units. This is shown in subsection (d) of the same column. With this the self-synchronization of the gait pattern by the robot's legs ends, and the robot continues to walk with a slower pace due to the decreased length of the swing and stance phases.

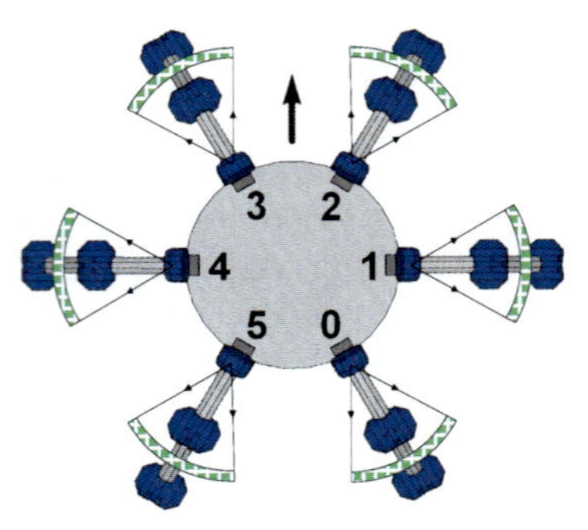

Fig. 5.13 Model of a hexapod robot with its legs numbered. The arrow represents the front of the robot. The arcs represent how lengthy the swing/stance phase is for each of the legs, which is initially the same for all legs but changes with gait pattern length change / self-synchronization.

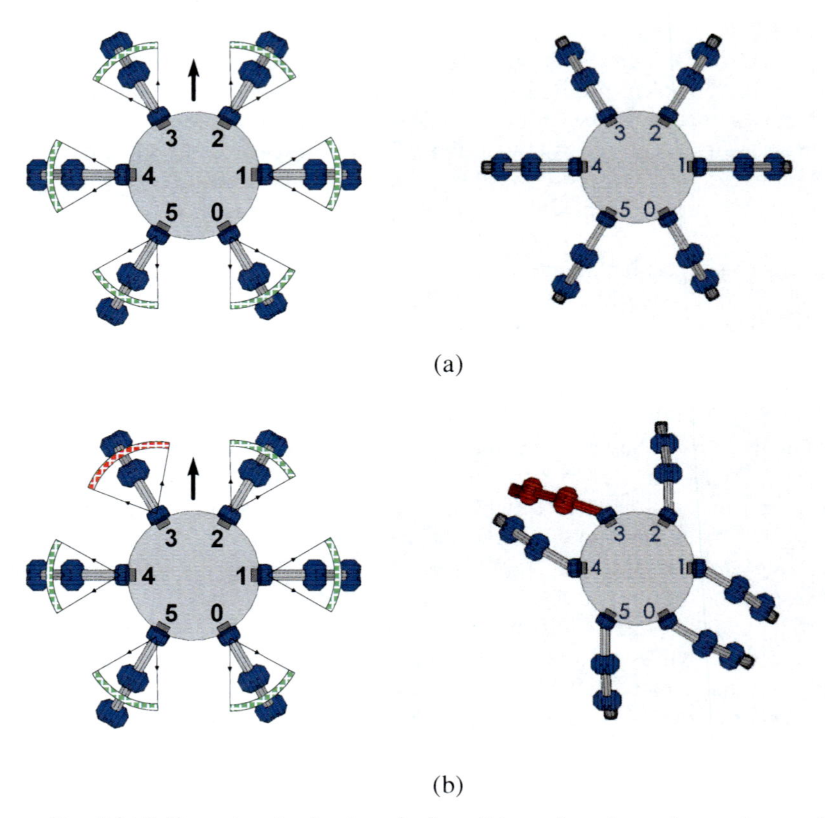

Fig. 5.14 Self-synchronization by robot's walking gait prolongation – robot model.

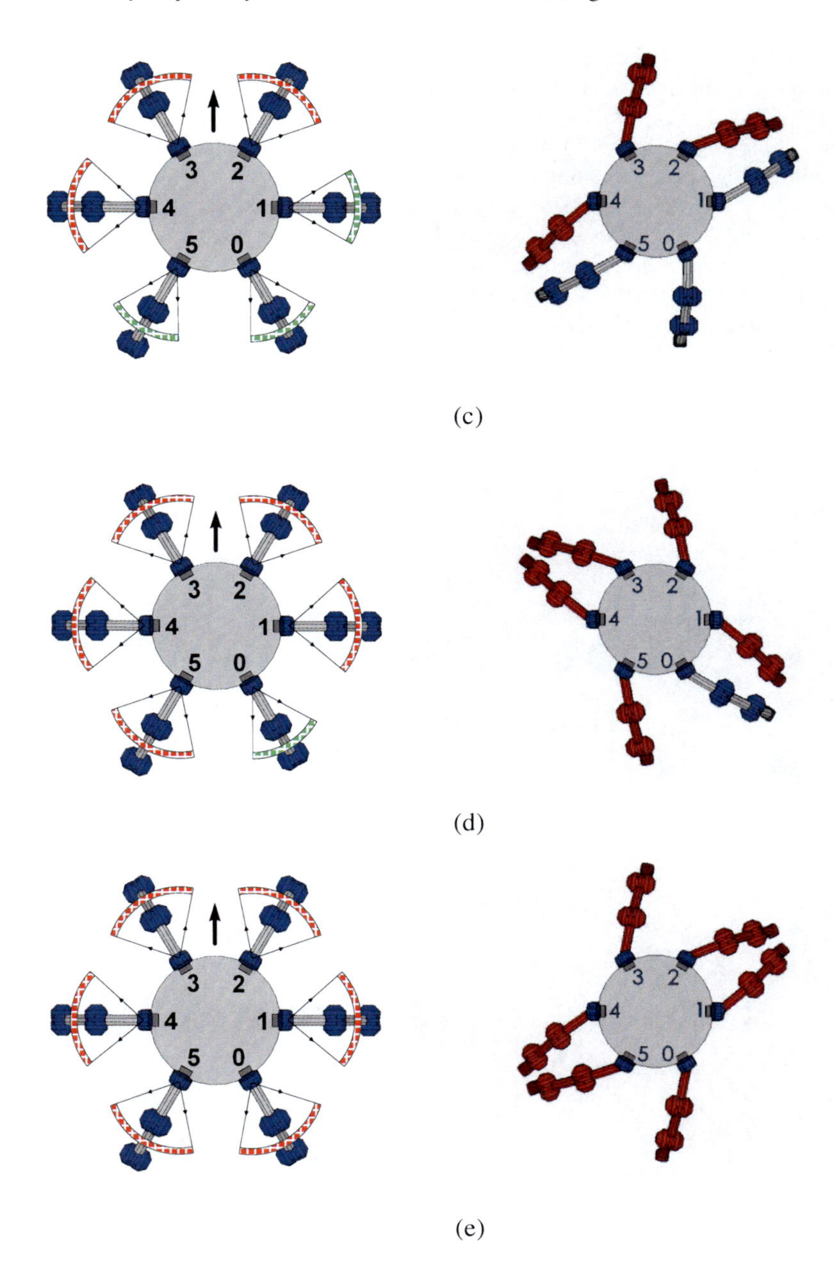

(c)

(d)

(e)

Fig. 5.14 (*continued*)

The previously explained self-synchronization of a robot's walking gait by prolongation and shortening is represented using the robot model, in Figure 5.14 (a), (b), (c), (d) and in Figure 5.15 (a), (b), (c), (d). The arcs over the legs represent the radius of leg movement. Differently colored arcs represent the increase or decrease of the leg operational angles during synchronization.

In Figure 5.14 (a), on the left side the robot model represents the situation where the legs are in their initial stance/swing lengths. On the right side the hexapod robot model is shown in its initial situation.

In Figure 5.14 (b), on the left side the robot model represents the situation where leg number 3 prolongs its stance phase from 3 length units to 4 length units. On the right side the hexapod robot model is represented where leg number 3 (indicated with red) prolongs its stance phase. This figure illustrates the situation in Figure 5.12 (a) - column "Synchro By Prolongation".

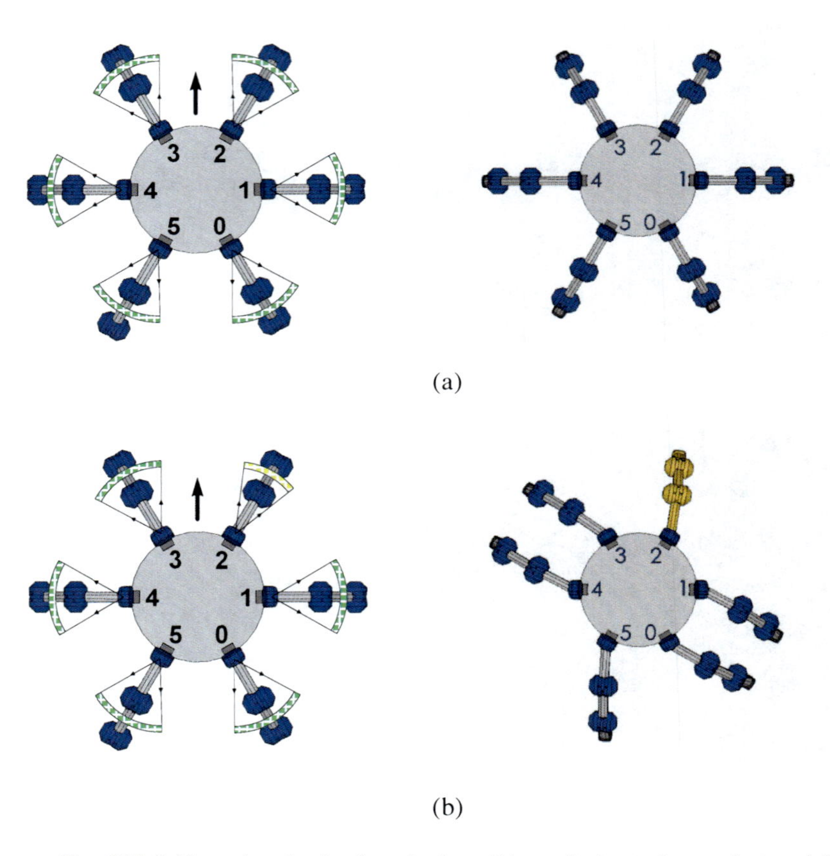

Fig. 5.15 Self-synchronization by robot's walking gait shortening – robot model.

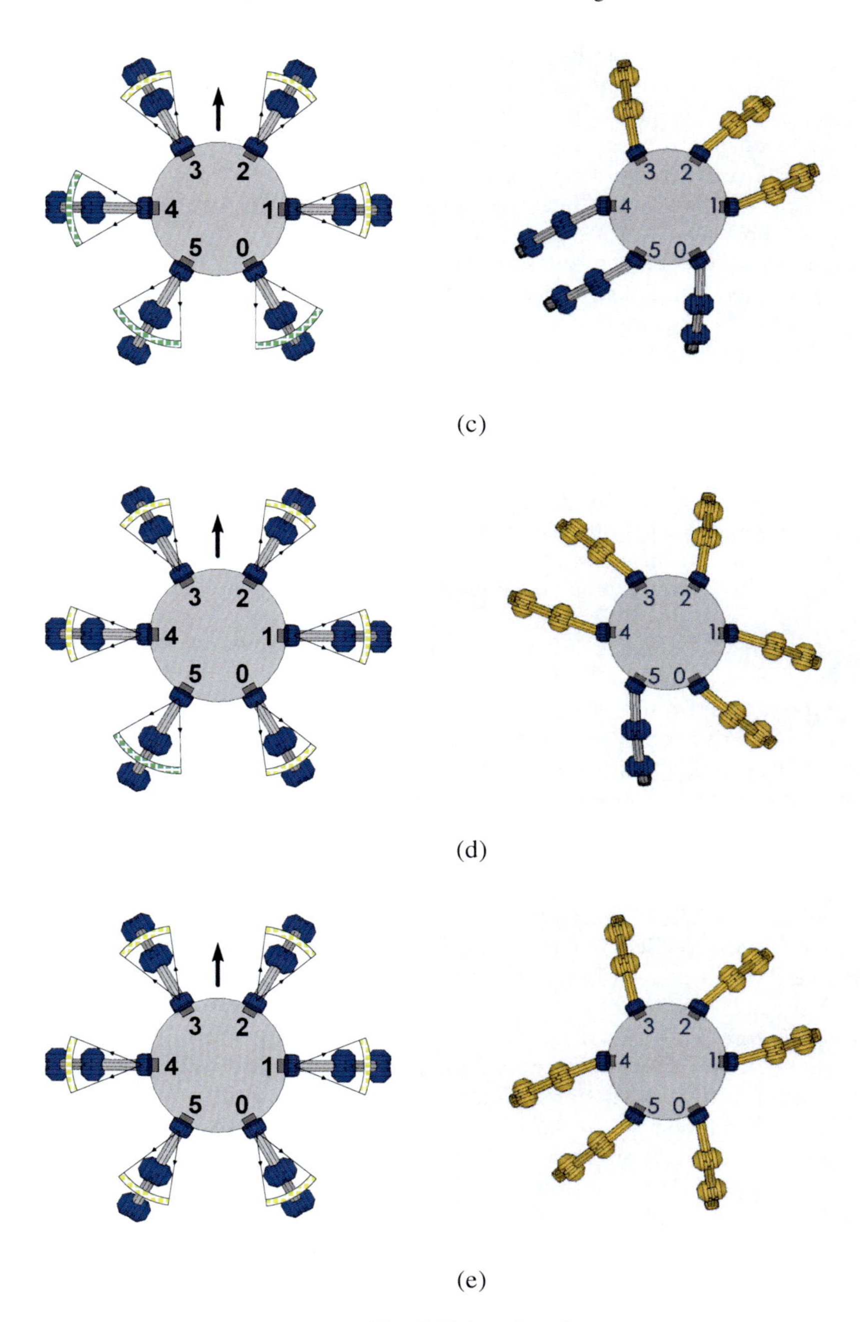

Fig. 5.15 (*continued*)

In Figure 5.14 (c), on the left side the robot model represents the situation where the neighboring legs numbered 2 and 4 synchronize their gait length in their stance phase from 3 length units to the length of the gait of leg 3: 4 length units. On the right side is the hexapod robot model where legs numbered 2 and 4 (indicated with red) prolong their stance phase and synchronize their gait length with leg 3. This figure illustrates the situation in Figure 5.12 (b) - column "Synchro By Prolongation".

In Figure 5.14 (d), on the left side the robot model represents the situation where the neighboring legs numbered 1 and 5 synchronize their gait length from 3 length units to 4 length units during their stance phase. On the right side is the hexapod robot model where legs numbered 1 and 5 (indicated with red) prolong their stance phase to synchronize their gait length. This figure illustrates the situation in Figure 5.12 (c) - column "Synchro By Prolongation".

In Figure 5.14 (c), on the left side the robot model represents situation where the last not-synchronized leg numbered 0 synchronizes its gait length from 3 length units to 4 length units during its stance phase. On the right side is the hexapod robot model where the leg numbered 0 (indicated with red) prolongs its stance phase and synchronizes its gait length with the neighboring legs. This figure illustrates the situation in Figure 5.12 (d) - column "Synchro By Prolongation".

In Figure 5.15 (a), on the left side the robot model represents the situation where the legs are in their initial stance/swing lengths. On the right side the hexapod robot model is shown in its initial situation.

In Figure 5.15 (b), the left side model represents the situation where leg number 2 decreases its gait length in the swing phase from 3 length units to 2 length units. On the right side the hexapod robot model is shown with leg number 2 (indicated with yellow color) shortening its swing phase. This figure illustrates the situation in Figure 5.12 (a) in the column "Synchro By Shortening".

In Figure 5.15 (c), on the left side the robot model represents the situation where the legs numbered 1 and 3 decrease their gait lengths in their swing phases from 3 length units to 2 length units. On the right side the hexapod robot model is shown where legs 1 and 3 (indicated with yellow) shorten their swing phase and synchronize their gait length with the other legs. This figure illustrates the situation in Figure 5.12 (b) in the column "Synchro By Shortening".

In Figure 5.15 (d), the left side robot model represents the situation where the legs numbered 0 and 4 shorten their gait length from 3 length units to 2 length units. On the right side the hexapod robot model is shown where legs 0 and 4 (indicated with yellow) shorten their swing phase and synchronize their gait length with other legs. This figure illustrates the situation in Figure 5.12 (c) in the column "Synchro By Shortening".

In Figure 5.15 (e), on the left side the robot model represents the situation where the last not synchronized leg, number 5, is synchronizing and shortens its gait length from 3 length units to 2 length units. On the right side the hexapod robot model is shown where leg 5 (indicated with yellow) shortens its swing phase and synchronizes its gait length with the neighboring legs. This figure illustrates the situation in Figure 5.12 (d) - column "Synchro By Shortening".

The overall *self-synchronization* by prolongation of gait pattern or by shortening of the length of gait pattern depends on the final gait length units to which the legs synchronize their swing and stance phases. On other side, the stability of the whole synchronization process in sense of how oscillating it is depends on how often such synchronization gets started by some external influences in some rather short time domain.

5.5 Implementation of Firefly Inspired Self-synchronization into the Robot Control Architecture

For demonstration of the practical usefulness of this approach, the experimental approach has been tested on the hexapod robot OSCAR-X. Additionally a distributed ORCA architecture (explained in chapter 4.4) is used in this approach.

The distributed control architecture for this experiment consists of the following:

- One OCU (Organic Control Unit) – related to monitoring of the proper function of the BCU units and to provide counteractions in case of anomalies.

The robot's architecture has decentralized organization where each of the six legs of the robot's control architecture consists of:

- One BCU related to leg gait generation (swing, stance phase) and firefly inspired synchronization;
- Three BCUs representing the three servo segments of each leg.

In such control architecture, the BCU related to gait generation communicates with its neighboring BCU units (also related to gait generation) and in that way provides means for the firefly self-synchronization mechanism which was explained earlier. The other 3 BCUs are related to servo positioning and control.

The decentralized architecture is also useful for generation of a self-organizing emergent walking gait as described in chapter 5.3. Therefore the robot can exhibit self-synchronization of the length of its walking gait pattern and at the same time continue walking with a self-organized walking pattern.

While this may prove to be very useful for developing a robust walking robot platform, describing in detail and visually presenting the whole process that happens during the walking gait self-synchronization and self-organization perhaps is too complex. The complexity originates from tracking the firefly inspired self-synchronization of the walking gait pattern and at the same time observing the swing/stance phases, which may or may not give immediate explanation about the process of self-synchronization depending on the reader.

In order to mitigate this and also present an example for the mechanism of self-synchronization, a series of experiments conducted on OSCAR-X were done. The walking gait pattern is selected to be *tripod gait* (explained in chapter 5.2.1) which is the fastest gait found in insects and also often used in the domain of walking robots. Instead of using the self-organizing walking gait, the tripod walking gait is used by the robot. The tripod gait for self-synchronization testing purposes can be also implemented (besides using the CPG – explained in chapter 5.2) by preprogramming the two groups of three legs that exchange their swing and stance patterns in the time domain. Choosing such a *tripod* gait helps the robot's walking gait to be clearly characterized and at the same time able to show the principle of firefly self-synchronization of walking robot gait. Additionally, by using some other walking gait aside from the self-organizing gait, it can also be shown that self-synchronization of the length of a walking gait can also be done by using other types of walking robot gaits.

5.6 Experiments Done with Firefly Inspired Self-synchronization and Results from Experiments

The aim of the experiments was to show how the firefly self-synchronization happens by prolongation and shortening the walking gait. Additional tracking was done on how the robot keeps its direction of walking by such self-synchronization of the length of its walking gait. In order to concentrate on these aspects, the tripod gait was chosen for reasons previously explained.

The stance and swing phases of a robot's walking gait are characterized by their length or as the angular path projected on the ground.

In the experiments done, the length unit 1 of swing / stance phase corresponds to angular movement of 10 degrees projected on the ground. The length unit of 2 corresponds to angular movement of 20 degrees. Analogously the length units of 3, 4 and 5 correspond to angular movement of 30, 40 and 50 degrees, respectively.

After the legs are all synchronized, some predefined legs of the robot suddenly increase or decrease their swing or stance length. The change of length of the leg's stance or swing gait simulates situations where the leg hits some object on the way and has to increase its stance / swing phase in order to overcome that object, or perhaps the leg has to decrease its swing / stance phase in order to walk better over some sandy terrain. The change of the length of swing / stance phase of one leg is followed by a firefly inspired self-synchronization of the gait lengths of the other robot's legs until they reach the same length of their swing/stance phases. This process is repeated when other legs start to prolong or shorten the length of their swing / stance phases.

The tracking of the robot during the experiment was done with the tracking setup as explained in the Appendix, Chapter 11.1- Test bed for tracking the robot OSCAR-X during the experiments.

There were three test cases and experiments done:

- Experiment about self-synchronization by prolongation of the robot's swing and stance phases from length 1 to length 2 to length 3. Legs 3 and 1 (Figure 5.16) on the robot are chosen to increase the length of their swing / stance phases. The experiment involves tracking the robot's path on the ground during the self-synchronization;

- Experiment about self-synchronization by shortening of the robot's swing and stance phases from length 5 to length 4 to length 3. Legs 2 and 4 (Figure 5.16) are chosen to decrease the length of their swing / stance phases. The experiment involves tracking the robot's path on the ground during the self-synchronization;
- Experiment about self-synchronization by combined prolongation and shortening of the robot's swing and stance phases from length 1 to length 2 to length 3 and then shortening from length 3 to length 2 to length 1. Legs 5 and 2 (Figure 5.16) are chosen to increase the length of their swing / stance phases. Legs 1 and 4 (Figure 5.16) are chosen to decrease the length of their swing / stance phases. The experiment involves tracking the robot's path on the ground during the self-synchronization;

Figures shown in the following sub-chapters give an overview on the experiments done with self-synchronization by consecutive prolongation, shortening, or a combination of both for the leg's swing/stance phases. The red circles indicate the moment when prolongation happens, by which leg and in which phase of the movement of the leg – stance or swing. The horizontal axis indicates the time in seconds and the vertical axis indicates the swing and stance phase ("SW" for swing phase and "ST" for stance phase) for each of the robot's legs from 0 through 5 (Figure 5.16). Each leg has a different representation pattern on the figure. A symbol at a particular point in time by a particular leg in row "SW" indicates that the leg is in its swing phase. A symbol at a particular point in time by a particular leg in row "ST" indicates that the leg is in its stance phase.

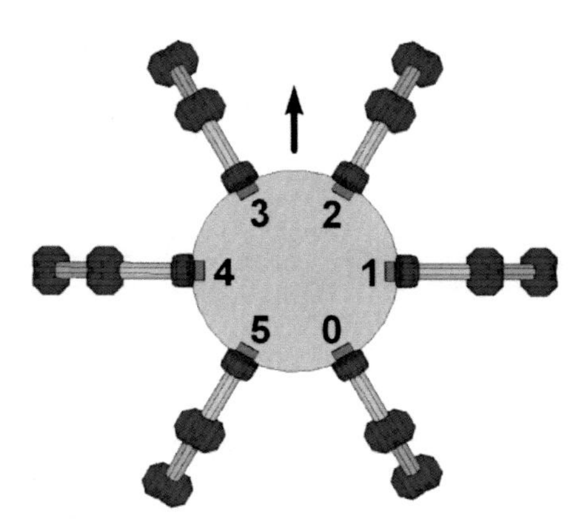

Fig. 5.16 Model of a hexapod robot with its legs numbered.

5.6.1 Experiment about Self-synchronization by Prolongation of the Robot's Swing and Stance Phases

In this experiment, the robot OSCAR-X first starts walking with a parameter length of 1 for the swing and stance phases of the robot's legs. With consecutive initialized prolongations of stance / swing phases by the robot's legs, the robot prolongs the length of swing / stance phase first from length 1 to length 2 in leg 3, then self-synchronization of the length walking gait pattern of the other legs takes place. After that, leg 1 increases the length of its swing / stance parameter from length 2 to length 3. After that the self-synchronization of the length walking gait pattern takes place in the other legs. At the end of the experiment all legs have a length of 3 for their swing / stance phases. All of this is shown in Figure 5.17.

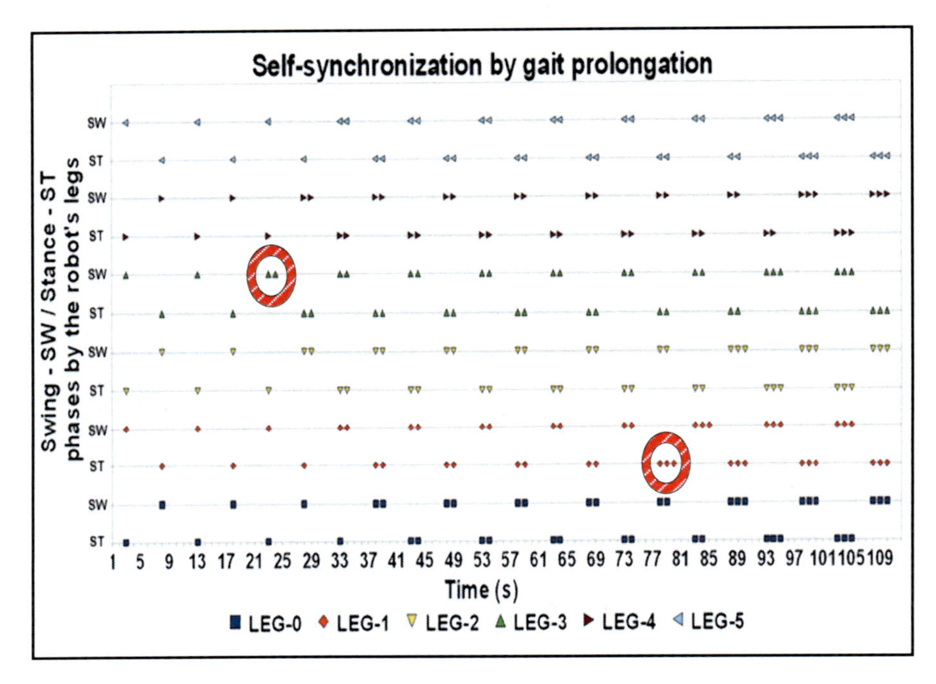

Fig. 5.17 Self-synchronization by prolongation of the robot's swing and stance phases.

Here are the details of the experiment:

First the length parameter is changed for leg 3 at 23s during its stance phase from length unit 1 to length unit 2. Then in the next cycle the neighboring legs 2 and 4 in their stance phases prolong their length parameters at time 28s. At time 33s, legs 1 and 5 also synchronize their swing and stance length units to 2 in their stance phases. At time 38s leg 0 also synchronizes its length of swing and stance to 2 length units. After this, all legs have the length of their swing / stance phases at 2 and self-synchronization is finished. In a similar fashion, the self-synchronization and prolongation of stance and swing phases is further done for each of the legs prolonging their phases from 2 length units to 3 length units. First leg 1 at 80s prolongs its swing / stance parameter from length 2 to length 3. After this legs 0 and 2 also synchronize their length to 3 units at 89s. Following this, legs 5 and 3 synchronize the length of their swing/stance phases to 3 units at 94s. Lastly, leg 4 adapts its length of the swing / stance phase to 3 units at 100s. With this the self-synchronization is finished and all of the legs have the same length unit of their swing / stance phases.

The tracking of the robot while performing the self-synchronization in this experiment is shown in Figure 5.18. The horizontal and vertical axes are in cm. The robot self-synchronization starts tracking at a point around 110cm on the horizontal axis and around 60cm on the vertical axis. The tracking ends at a point with coordinates around 215cm on the horizontal axis and 90cm on the vertical axis. From this it can easily be seen that the robot turns right while walking instead of keeping its original heading which is a straight vector parallel with the vertical axis and crossing the horizontal axis at 110 cm.

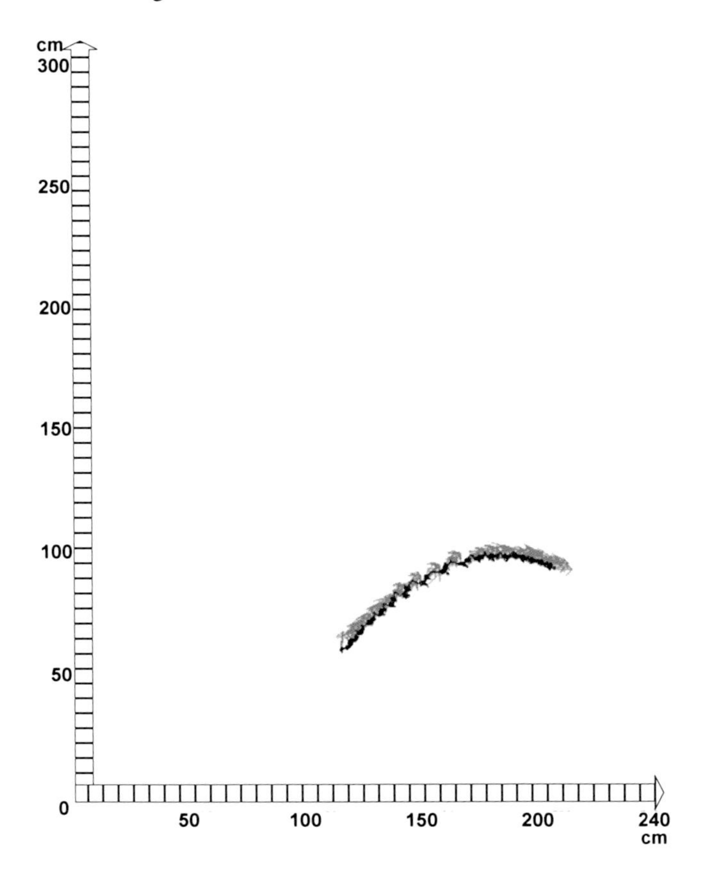

Fig. 5.18 Tracking of the robot during the self-synchronization by prolongation of its swing and stance phases.

This deviation from the original course can be related to the prolongation of the stance / swing phases first in the left group of legs which influences the turning rate, and also depends on the walking dynamics of the robot over the terrain. The tracing is done for pure forward motion behavior of the robot without any turning behaviors, so the influence of the self-synchronization on the robot walking heading can be more easily seen. Knowing the influence of the self-synchronization of robot's walking gait pattern on the robot's heading can help in choosing the turning robot behaviors which compensate for heading deviations, so the robot always keeps its proposed heading.

5.6.2 Experiment about Self-synchronization by Shortening of the Robot's Swing and Stance Phases

In the second experiment the robot OSCAR-X starts walking with a parameter length of 5 for the swing/stance phases of the robot's legs. With consecutive initialized shortening of stance / swing phases by robot's legs, the robot shortens the length of swing / stance phases first from length 5 to length 4 for leg 2, then self-synchronization of the length walking gait pattern takes place for the other legs. After that, leg 4, decreases the length of its swing / stance parameter from length 4 to length 3. Then the self-synchronization of the length walking gait pattern takes place for the other legs. At the end of the experiment all legs have a length of 3 for their swing / stance phases. All this is presented in Figure 5.19.

Here are the details of the experiment:

The robot starts to walk with a length parameter of 5 for the swing and stance phases. First at 17s leg 2 decreases its swing phase the swing / stance parameter from 5 length units to 4 length units. In the second cycle legs 1 and 3 decrease their stance / swing parameters in their swing phases to length of 4 units at 22s. Then the synchronization in the swing phase by legs 0 and 4 to 4 length units follows at 27s. At 32s leg number 5 also synchronizes its stance / swing length parameter to a length of 4. In similar fashion, leg 4 decreases the length of its swing / stance phase from 4 on 3 unit lengths at 69s.

Decreasing of the swing / stance walking length parameters from 4 length units to 3 length units for legs 5 and 3 follows at 79s. Self-synchronization continues with adaptation of the length of swing / stance phases for legs 0 and 2 at 83s. At the end leg 1 also synchronizes its length of swing / stance phase to 3 length units at 88s. Self-synchronization also continues for stance / swing lengths until 87s when all the legs have the length of the stance / swing parameter of 3. With this the self-synchronization is finished and the legs all have the same length unit for their swing / stance phases.

In Figure 5.20 the tracking of the robot is shown during performing this experiment. The horizontal and vertical axes are in centimeters. The robot self-synchronization starts tracking at a point around 110cm on the horizontal axis and around 60cm on the vertical axis. The tracking ends at a point with coordinates around 150cm on the horizontal axis and 180cm on the vertical axis. It can be seen that the robot does not deviate too much from its original heading which is a straight vector parallel with the vertical axis and crossing the horizontal axis at 110 cm. A small deviation from the main heading is still present however. This is most likely due to the influence of the legs when they start changing the length of their swing / stance phases, and then are followed by self-synchronization of the length of swing / stance phases by the other robot's legs. Since the experiment was done on shortening the length of swing / stance phases, the decrease in speed over the terrain (decreasing the length of swing / stance phases has a direct influence on the speed of the robot) probably has less impact on the change in direction of the robot than during the self-synchronization by prolongation of robot's walking gaits.

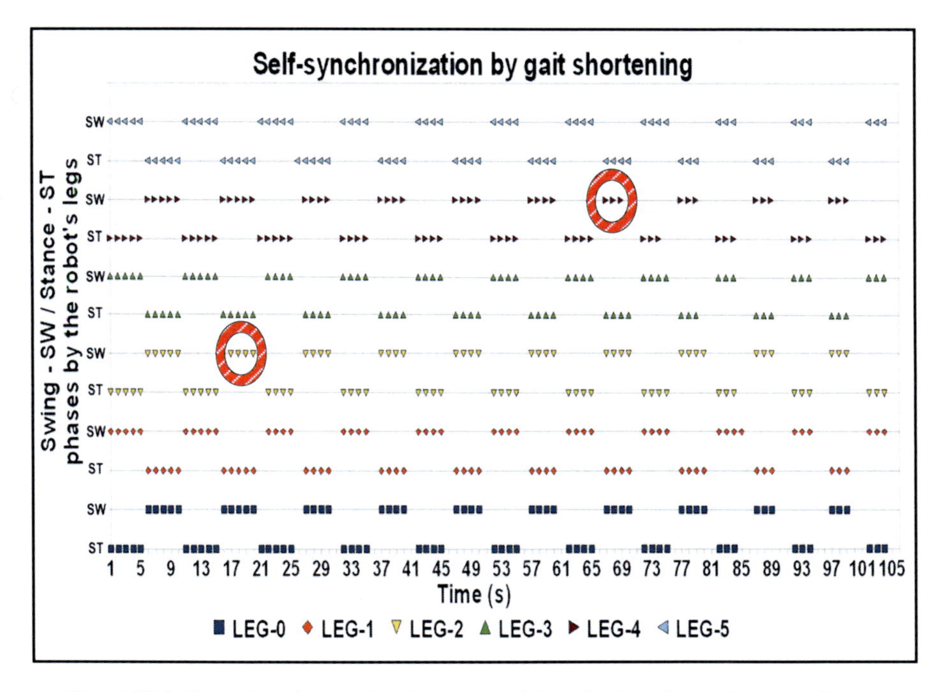

Fig. 5.19 Self-synchronization by shortening of the robot's swing and stance phases.

The tracking is done by forward going behavior of the robot without any side behavior turnings, so the influence of the self-synchronization on robot's walking heading can be more accurately observed. In this experiment the robot doesn't deviate much from the original straightforward heading, however knowing the influence of the self-synchronization of robot walking gait pattern on the heading of the robot can help in choosing the robot turning behaviors to compensate and keep the robot on its proposed heading.

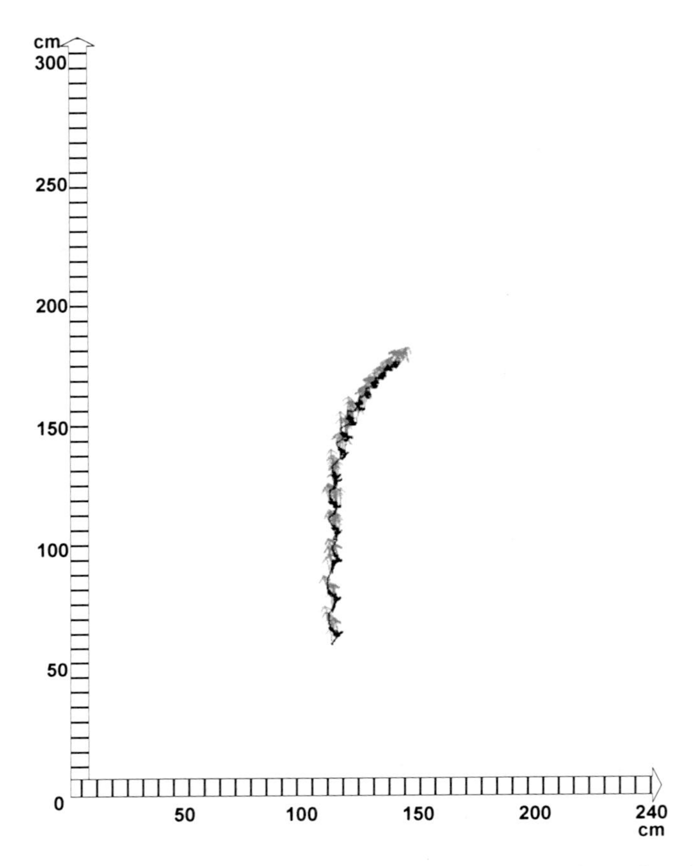

Fig. 5.20 Tracking of the robot during the self-synchronization by shortening of its swing and stance phases.

5.6.3 Experiment about Self-synchronization by Combined Prolongation and Shortening of the Robot's Swing and Stance Phases

In the third experiment the combined scenario is tested with prolongation followed by shortening of the swing/stance phases by robot's walking gait. Namely, robot OSCAR-X starts walking with a parameter length of 5 for the swing/stance phases by the robot's legs. With consecutive initialized prolongation of stance / swing phases of the robot's legs, the robot prolongs the length of swing / stance phase first from length 1 to length 2 in leg 5, then self-synchronization of the length walking gait pattern takes place by the other legs. After that, leg 2 increases the length of its swing / stance parameter from length 2 to length 3. After that the self-synchronization of the length walking gait pattern takes place by the other legs. After prolongation a series of shortening of the swing / stance phase lengths of the robot's legs follows. With consecutive shortening of stance / swing phases by robot's legs, the robot shortens the length of swing / stance phase first from length 3 to length 2 in leg 1, then self-synchronization of the length walking gait pattern takes place for the other legs. After that, leg 4 decreases the length of its swing / stance parameter from length 2 to length 1. Then the self-synchronization of the walking gait length takes place by the other legs. At the start of the experiment all legs have a length of 1 for their swing / stance phases and at the end of the experiment all legs are back to length of 1 again for their phases. All this is presented in Figure 5.21.

Here are the details of the experiment:

In this experiment a dynamic self-synchronization of the robot walking gait pattern is performed, starting from a stance / swing parameter length of 1, then increasing to a parameter length of 3, and finally decreasing again to a stance / swing parameter length of 1. The first change of the parameter occurs in leg 5 within its stance phase at 20s. After this cycle, the next parameter change occurs by neighboring legs 0 and 4 in their stance phases at about 30s, increasing from a length parameter of 1 to a length parameter of 2. In the next cycle the legs numbered 1 and 3 get synchronized to parameter length of 2 at about 35s. The synchronization continues within leg number 2, which adjusts its parameter from length 1 to length 2 at 40s.

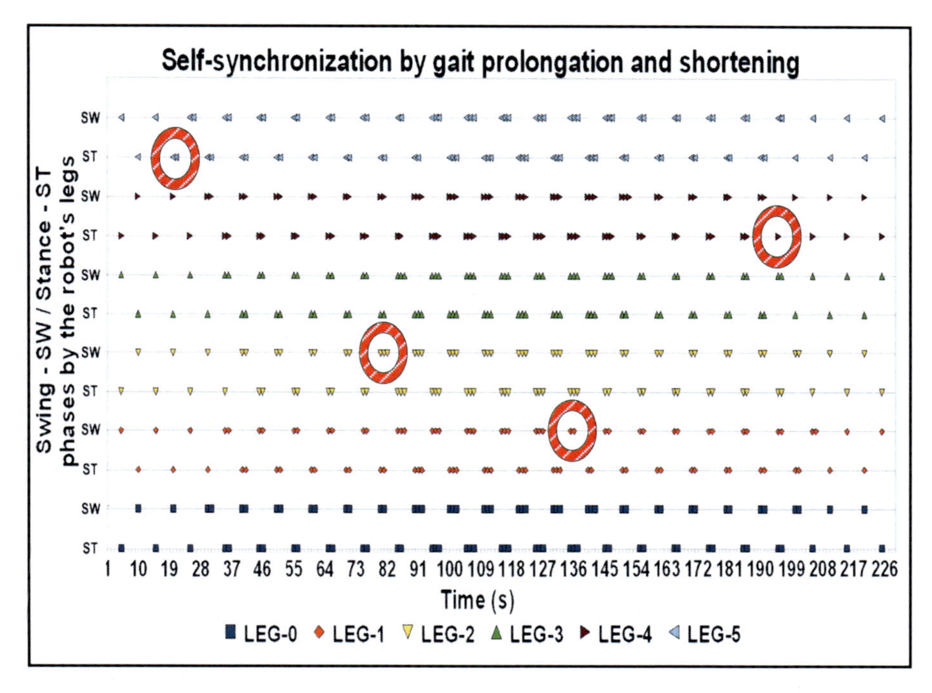

Fig. 5.21 Self-synchronization by prolongation and shortening of the robot's swing and stance phases.

In a similar fashion, the self-synchronization continues in the same experiment also for prolongation of the stance / swing phases from length parameter 2 to length parameter 3 for each of the legs. At 81s, leg 2 increases its length of swing / stance parameter from 2 to 3 length units. Then an increase of length swing / stance parameter from 2 to 3 units by legs 3 and 1 follows at about 85s. At 91s, legs 4 and 0 increase the length of their swing / stance parameter from 2 to 3 length units to synchronize their swing / stance parameters. Around 96s, leg 5 synchronizes its swing / stance length from 2 to 3 length units. With this the self-synchronization cycle is finished and all legs have same length for their swing / stance parameter.

The shortening of stance / swing parameters takes place similar to the prolongation in this experiment. At 136s, leg 1 decreases its stance / swing length parameter in its swing phase from 3 to 2 length units. After this the legs numbered 0 and 2 also decrease their swing / stance parameter from 3 to 2 length units at 145s. A decrease of swing / stance length parameter from 3 to 2 unit length by legs numbered 5 and 3 follows at 150s. At the end leg 4 also decreases its swing / stance length parameter from 3 to 2 units at 159s. With this the self-synchronization cycle is finished and all the legs have a swing / stance length parameter of 2 length units. In similar fashion further decrease of swing / stance length parameter takes place. Leg 4 decreases its swing / stance phase from 2 to 1 unit length at 198s. With the self-synchronization principle, a decrease of the swings / stance length parameters of legs 5 and 3 from 2 to 1 unit length occurs at 203s. After this cycle the next decreasing of swing / stance length parameters takes place by legs 0 and 2 from 2 to 1 unit length at 207s. Finally leg number 1 also decreases its swing / stance length from 2 to 1 unit length at 212s. With this the self-synchronization process is finished.

This last experiment demonstrates that dynamic self-synchronization by increasing and decreasing of the gait parameters can be performed one after another during the walking of the robot.

In Figure 5.22 the tracking of the robot is shown during performing the self-synchronization by consecutive prolongation and shortening of the swing / stance phases. The horizontal and vertical axes are in centimeters. The robot self-synchronization starts to be tracked at a point around 110cm on the horizontal axis and around 60cm on the vertical axis. The tracking ends at a point with coordinates around 190cm on the horizontal axis and 190cm on the vertical axis. The robot's path shows that during the self-synchronization by prolongation in its first phase of walking the robot has a tendency to turn slightly toward the right. This is probably an effect of prolongation of the leg's swing / stance phase which somehow influences the robot to turn.

This is due to the unequal transient swing / stance lengths between the legs, which leads the robot to start to turn from its original heading. Therefore the path of the first phase of robot's walking is similar to the experiment done with self-synchronization only by prolongation of the robot's legs swing / stance phases. In the second phase of the robot's walking, it can be seen that the robot starts to turn slightly toward the left. The second phase of the robot's walking in this experiment is associated with the self-synchronization by shortening of the robot's walking gait. The shortening of the robot's swing / stance parameters probably influences the robot's walking to turn slightly toward the left. The tracking is done with forward like movement behavior of the robot without any side turning behaviors, so the influence of the self-synchronization on the robot's heading can be more accurately tracked. The parameters of the swing / stance phases change by the self-synchronization process from length 1 to length 3 units and then back again decreasing from 3 units to 1. In this experiment the robot's path first slightly turns toward the right and then toward the left. This experiment can further help with developing and choosing the turning robot behaviors which help compensate these heading deviations so the robot always keeps its proposed heading.

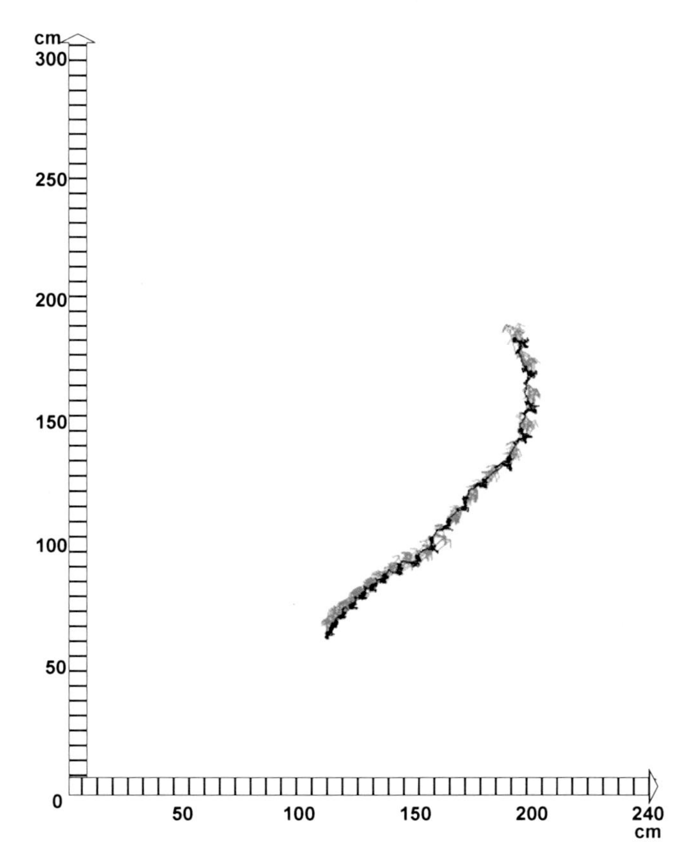

Fig. 5.22 Tracking of the robot during the self-synchronization by *prolongation and shortening* of its swing and stance phases.

5.6.4 Discussion on Future Possible Improvements of Firefly Inspired Self-synchronization Approach

There are some notions that would perhaps be interesting to be explored in the future for such firefly inspired self-synchronization of robot walking gait.

For example, stability of the self-synchronization approach can be further explored. Namely the stability of the whole self-synchronization process depends on how often such synchronization initializes due to some external influences in a rather short time domain. The shorter the period in which different synchronizations get initialized (different legs start to change their swing / stance length parameters) introduces divergence in the self-synchronization process. On the other hand, the process of self-synchronization is emergent and by emergence would converge towards self-synchronized walking robot gait. Therefore this idea can be further explored in future research.

The tracking experiments have shown that the robot by straight forward walking and by doing self-synchronization of its walking pattern may start to change its heading due to dissimilar length of swing / stance phases that the legs have by such a self-synchronization process. Future research may be done on implementing additional robot behaviors such as left and right turning that will be integrated in the decentralized robot control architecture and will compensate for the heading deviations so the robot will always keep the proposed heading.

5.6.5 Summary about the Firefly Inspired Self-synchronization Approach

In this chapter notions about firefly inspired *self-synchronization* for robot walking gait have been introduced and results from real scenario experiments done on the hexapod robot have been represented. By implementation of such firefly inspired self-synchronization and self-gait organizing walking, a versatile robot walking platform can be implemented that can change its walking gait pattern in a distributive and *self-organizing* way. In general, the firefly self-synchronization concept presented here can also be useful in other types of walking robot projects.

The decentralized self-organization of a robot's walking gait is also an interesting subject to be explored for the domain of reconfigurable walking robots (concept for self-reconfigurable walking robot is explained in one of the following chapters) and practically implemented within a decentralized robot control architecture. The possible application would consider synchronizing the walking gait of the robot after reconfiguration by the robot is performed (one of its legs is amputated) or for a scenario where the walking robot goes over an obstacle and in such cases the walking gait can be again re-synchronized by using a firefly inspired self-synchronization.

Chapter 6
Biologically Inspired Approach for Optimizing the Walking Gait of a Humanoid Robot

Humanoid robots are already present in many aspects of our lives: as service robots, as entertainment robots playing soccer, or simply as toys. They have been used as research subjects for testing our understanding of human motion and as test-beds for developing and testing of various locomotion algorithms.

However, since they are multi degrees of freedom (DOF) robots, having many joints, hardware, and software modules, etc. [KSW05] [TsL06] their programming is often not straight forward and involves a lot of rethinking on finding a stable locomotion and lowering their energy consumption. This is an essential problem also for the robots that play soccer in RoboCup matches [Rob10]. Namely, the difference between winning and losing the match is not just related to vision processing, ball shooting, and mapping tasks, but also on how fast the robot is, how stable it is, how often it falls to the ground or how often it has to be out of the game for replacing its batteries.

The research presented here addresses exactly these problems –how to develop a biologically inspired robust approach that will allow the humanoid robots to online self-optimize their walking gaits in respect to the robot's stability, walking speed, and lowering the energy consumption. The developed approach was intended to be used for humanoid robots walking on flat surfaces (concrete, linoleum, various types of carpets, etc.) representing the calibration phase in which the humanoid robot finds the optimal parameters for its walking [JKH10] [Kot09].

The usefulness of this approach was practically demonstrated in various experiments done under real conditions on the humanoid robot demonstrator S2-HuRo (described in Chapter 3.5).

Before describing the biologically inspired approach for humanoid robot self-stabilization in more detail, first an overview will be given about the motivation behind using such a biologically inspired approach.

6.1 Approaches for Walking Gait Generation by Humanoid Robots

Different approaches have been considered for biped robot locomotion [VBS90] [YaL03]. Some of them are based on mathematical models [VBS90] [YaL03] of the humanoid robots and describing the dynamic walking or the maintenance of Zero

B. Jakimovski: Biologically Inspired Approaches, COSMOS 14, pp. 67–125.
springerlink.com

Moment Pole (ZMP) inside the support region [VuJ69] [Vuk73] [LSY07] [CZT07]. Besides that, the full mathematical modeling of the humanoid robot can be cumbersome and error prone. The modeling of humanoid robots also induces another characteristic: the developed mathematical model for one type of robot is on the other hand most likely not transferable to any other type of humanoid robot. This means that everything has to be remodeled for such a particular humanoid robot so that the mathematically modeled approach can be also useful for that kind of robot.

Since the whole process of modeling can be rather burdensome, too time consuming, and at the end most likely not transferable to other type of robots, a biologically inspired approach has been chosen instead. This is due to the characteristics of biologically inspired approaches that introduce notions from biological systems and characteristics such as learning, self-optimization, and self-stabilization.

Different biologically inspired paradigms have been tried for the humanoid robot's domain [BaK06], [KKR06] [KKS06], [MEN08]. Some of them are based on spinal central pattern generators (CPGs) in vertebrate systems [BaK06], others use the CPG in relation with modulation of stiffness [KKR06], and reflex based stabilization uses SMA muscles [KKS06] or coupled oscillators [MEN08]. For humanoid robots numerous approaches of neural control architectures are known as well (Kuniyoshi, G. Cheng, Y. Nakamura,...). Results from those experiments have demonstrated that biologically inspired approaches can be very useful for the locomotion of humanoid robots, which was additional motivation for performing research on developing a biologically inspired approach for self-stabilizing humanoid robot.

The researchers often test the humanoid robot walking gait that they generate with their control algorithms in some simulation environments [QRY08], [FPV08], [CMZ08], [DWX08]. The benefit of using the simulations is that the mechanical integrity of the humanoid robots does not have to be sacrificed in case the control algorithm does not control the humanoid robot in the intended way. Often simulation environments are stated to provide high fidelity rigid body dynamics [Cyb09], [Pla09], [Msd09], [Usa09]. However, the simulation experiments cannot be completely identical with the reality experiments because of various factors such as: environmental influences, dynamics, vibrations, sensor noises, and other aspects present in reality. This may also imply that the tests done in simulation for particular control algorithms may give a false overview on how well they function, since direct implementation of the same algorithm tested in simulation may prove to control the real robot in a slightly different way. Therefore one may conclude that there perhaps isn't such "one-to-one" mapping from the control algorithms developed in the simulation and the same algorithm applied in real experiments. Since transferring of the algorithm will more or less rely on "tuning" the algorithm in order to be successfully used in the real conditions on a real robot.

For these reasons the biologically inspired approach for humanoid robot self-stabilization was tested solely under real conditions – without using any simulations, so that the approach developed can be guaranteed to function under real conditions on a real robot.

6.2 Symbiosis as a Biologically Inspired Approach for Self-stabilization of Humanoid Robot Walking Gait

The biologically inspired approach for self-stabilization of the walking gait of humanoid robot is called *SelSta* (from Self-Stabilization) and is based on the mutualistic Symbiosis processes [AhP00] seen in nature – which is a type of mutual interaction between biological species from which both species benefit. For example: Clownfish and sea anemone tentacles, coral organisms and algae, Goby fish and shrimp, Crabs and sea anemones, etc. In Clownfish and sea anemone symbiosis case, the symbiosis is such that the Clownfish feeds on the invertebrates which are dangerous for the sea anemone. Clownfish are the only fish that are able to live in sea anemones and not get stung by their tentacles. This is not the case for the other fish and therefore this provides sort of protection for Clownfish from other fish predators. And from other side the fecal matter from clownfish provides some essential nutrients for the sea anemone. This is how they benefit from living with each other in such symbiosis.

Fig. 6.1 Mutualism between Clown fish and sea anemone.

The Symbiosis mapped in *SelSta* is described by "mutual" interaction between the robot's lateral and longitudinal (or sagittal) axis stability and the load (level of current consumption) on the servos (Figure 6.2). This mutual interaction is directly related to the stability with which the robot walks over the ground and also the energy consumption (the less optimized servo movements consume more energy). The more optimized the interaction between the robot's axes, the more stable the robot is and less energy is consumed during walking gait generation. This is shown in Figure 6.3.

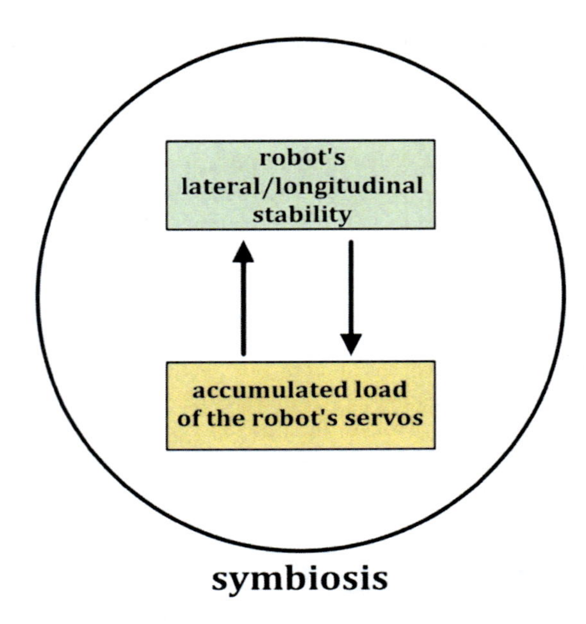

Fig. 6.2 Mutualistic symbiosis mapped in SelSta approach.

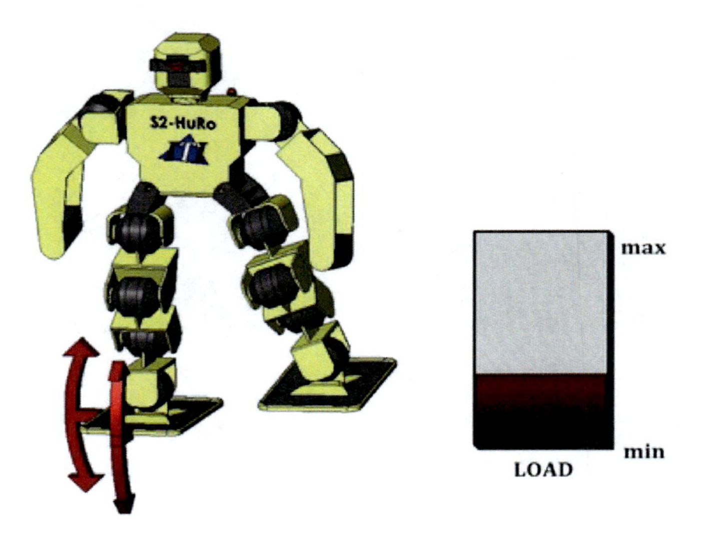

Fig. 6.3 "Mutual" interaction between the robot's lateral and longitudinal axes and its relation to robot stability and energy consumption level.

The *SelSta* approach is built as an add-on module for the already developed humanoid robot walking algorithm (not strictly optimized for walking on any particular surface) with a predefined non-optimal walking gait. The humanoid robot using SelSta first finds the best walking parameters in a calibration phase, so the robot can achieve the best performance in stability, speed, and energy consumption over a given flat surface. After the calibration phase with *SelSta*, the humanoid robot is then after only run with the best walking parameters found during the calibration phase.

6.3 SelSta Approach in Detail

Since the *SelSta* approach was designed to function under real circumstances on the humanoid robot S2-HuRo (described in Chapter 3.5), several details have to be considered for its practical implementation:

- S2-HuRo robot platform and sensors;
- Control of the robot S2-HuRo;
- Main parts of *SelSta* approach - *SymbScore* value and genetic algorithms;
- Fuzzy logic computation of *SymbScore* value;
- Genetic algorithm details for *SelSta* approach;
- Preparation for experiments;
- Experiments done with the *SelSta* approach.

Note: Some of the details explained here are related directly to the hardware of the S2-HuRo platform, for example: the types of gyros used, the embedded system, and load computation. However, the *SelSta* approach was made in a rather generalized way, so it can be transfered to different humanoid robot platforms and projects. The only thing that needs to be considered when transferring the *SelSta* to other humanoid robot platforms is obtaining the gyro values and load/energy consumption values which are then fed into the *SelSta* computation method.

6.3.1 S2-HuRo Humanoid Robot Platform and Sensors Used

The *SelSta* approach was tested on a S2-HuRo platform where the programs are run on the embedded system Gumstix® "Verdex board" [GUM09] with embedded Linux. The idea was that this approach should be demonstrated and run solely on the robot platform – therefore completely fulfilling the proposal to design a self-stabilizing walking humanoid robot. For the practical measurements and data logging a PC was used later, since the logging of data on the embedded system's MMC card takes a lot of time in comparison with the time for real time data logging on a PC. Also, a PC speeds up the evaluation process during the measurements done for humanoid robot self-stabilization. However, it was also considered that such a self-stabilizing process can run on the robot solely without any further need for additional external connections (or data logging) to PC for example.

There is also an ATmega 128 controller on the S2-HuRo robot, which is used to control the servos (using PWM signals), and also to acquire the dual axis gyro values (longitudinal, lateral) and feet sensor contacts (three sensor contact readings per leg foot - described in Chapter 3.5, Figure 3.15). The embedded system Gumstix® is connected to the onboard ATmega 128 controller though a serial line (reading the sensor values), and it runs a C program for sensor data interpretation, walking behaviors of the S2-HuRo like walk, stop, etc., as well as some other functions like standing up after a fall or commanded robot movements through a Wireless LAN (WLAN) connection, etc.

Figure 6.4 represents the used hardware.

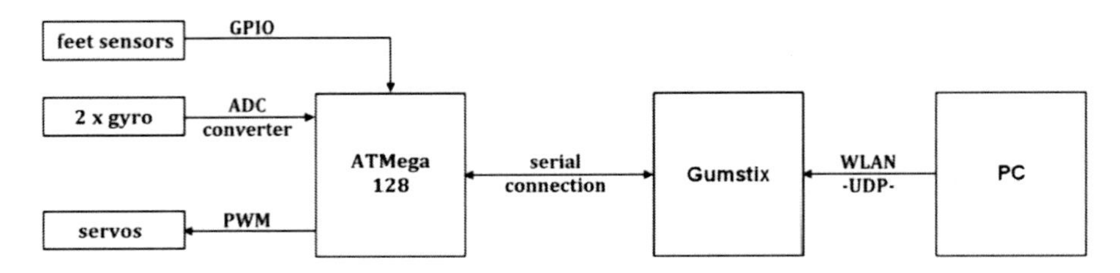

Fig. 6.4 Representation of the hardware setup used by the S2-HuRo and SelSta approach.

For the experiments done, the program running on Gumstix® was compiled for x86 architecture and executed on a PC instead, so the experiments and data logging can be done faster.

6.3.2 Control of the Robot S2-HuRo

For the SelSta experiments there are 3 robot walking modes considered:

- "Standard walking" - with standard swing / stance parameters for robot walking;

  This type of walking is represented by some pre-defined parameters for position of the legs/feet in each of the swing/stance phases with which the robot is walking more or less stably over a flat linoleum surface. The walking parameters by this walking are not optimized for stabilized humanoid robot walking, or for lowest energy consumption. This type of walking represents the walking of robots taken "out-of-the-box" or default programmed walking.
- "Manually optimized walking" - with manually optimized swing / stance parameters for robot walking on a particular flat surface;

  For this walking, the parameters for legs/feet positions by swing / stance are manually optimized for a flat linoleum surface, so the robot performs stabilized walking (however not necessarily the one with lowest energy consumption). This type of walking may take 10-12 human hours or even days before optimal walking parameters can be found. This type of walking represents the walking for humanoid robots for Robocup events [Rob10] that is optimized by human operators for a particular carpet surface for that event.

- *SelSta* optimized set of swing / stance parameters for robot walking on a particular flat surface;

  This type of humanoid robot walking is where the swing / stance parameters of the humanoid robot's walking are optimized using the *SelSta* approach, so the robot performs the most stable and least energy consuming walking on a particular flat surface. The *SelSta* approach starts with "Standard walking" and optimizes the walking parameters until optimal walking parameters are found.

For guaranteeing the consistency in measurements done with the *SelSta* approach, the movement of the S2-HuRo is done in 80ms intervals. Such interval based movement is more reliable when speaking about the precision of the movements, since in that given interval the servos are moved to the commanded position. The servos are also set to run with their highest speed. Without the interval there will be inconsistency in the movement of servos, for example a failure in some movements when some servos cannot achieve the commanded position before another command is given to move the servos to another position.

Given that the servos are updated every 80ms, and taking into account that the distance traversed by the leg on the ground is the same, the interval type of movement also gives a possibility that the speed of the robot can be controlled, i.e. the more walking intervals, the slower the speed of the robot since many leg position updates must take place. In a similar way, the smaller the number of walking intervals, the greater the ground speed of the robot over the terrain since leg positions of the robot are updated in fewer intervals.

For the experiments with the *SelSta* approach, the following speed categories have been considered:

- Slow speed - with 20 x 80ms walking intervals;
- Medium speed - with 15 x 80ms walking intervals;
- Fast speed - with 12 x 80ms walking intervals;

By movement of the robot's leg, the foot moves on the ground in its stance phase from anterior extreme position (AEP) to posterior extreme position (PEP), and in the air during its swing phase from the PEP position to AEP position. Where:

- AEP position is when the leg is at the end of its return stroke, or at the end of its swing phase;
- PEP position is when the leg is at the end of its power stroke, or at the end of its stance phase.

Each robot's leg movement can be represented as in the figure below.

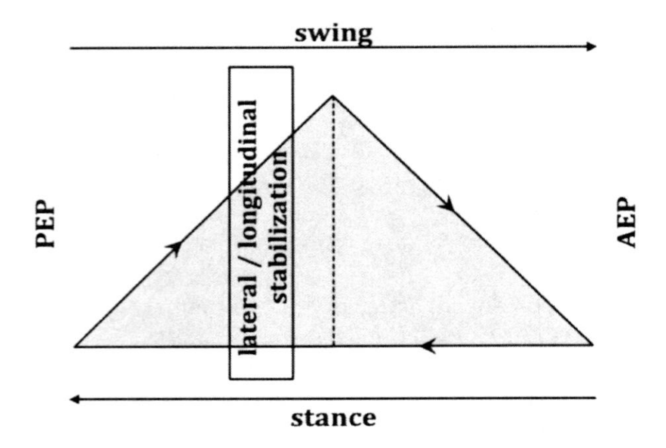

Fig. 6.5 Swing and stance phases by S2-HuRo leg. Lateral / longitudinal stabilization region represents where stabilization of the robot using *SelSta* approach takes place.

The *SelSta* approach considers the rectangular region indicated with "lateral/ longitudinal stabilization", where the stabilization takes part during each of the robot's walking steps.

6.3.3 Main Parts of SelSta Approach – SymbScore Value and Genetic Algorithm

As previously described, the *SelSta* approach is based on the biological paradigm of Symbiosis, which by the self-stabilization is interpreted as "mutual" interaction between the robot's lateral/longitudinal stability and the accumulated load of the robot's servos. Also "mutual" interaction can be found on another level - between the robot's lateral and longitudinal (or sagittal) axis stability. The For the *SelSta* approach, the load on the servos was also considered, which is directly related to current consumption of the servos during the robot's walking. The parameter *SymbScore* is introduced for quantitatively describing how good or bad the chosen parameters are for the robot's walking. The *SymbScore* value is in a range from 0 and 1, describing the stability of the robot for its walking, ranging from "non-stable" (value=0), up to "stable" walking (value=1).

The *SelSta* approach uses cascaded fuzzy logic for the *SymbScore* value computation and also a genetic algorithm to find the best possible walking parameters. This entails analyzing lateral and longitudinal foot and hip positions and their angles with respect to the ground such that the *SymbScore* value is maximized, i.e. the stability of the robot is maximized with the smallest energy consumption.

The main schematic for calculating the *SymbScore* value is shown in Figure 6.6.

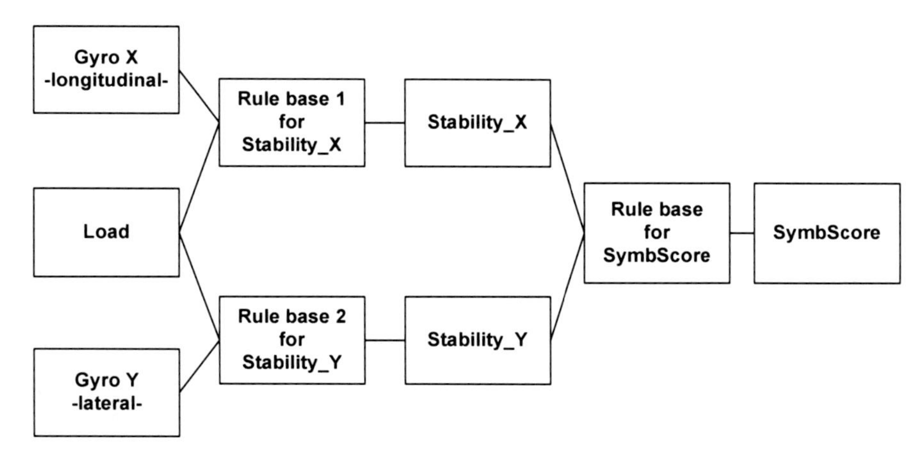

Fig. 6.6 Computation scheme for SymbScore.

As can be seen in the above figure, the computation of the *SymbScore* value consists of several intermediate steps. Firstly, on each axis the values for the gyros (lateral and longitudinal) and the accumulated load on all the robot's servos get fuzzified through the "rule base 1 for stabilty_X" and "rule base 2 for stability_Y" fuzzy logic rule bases, producing intermediate values named "Stability_X" and "Stability_Y". Hereafter, these two values get further fuzzified into the rule base for "symbiosis rule base for SymbScore" and with that the *SymbScore* value is computed.

6.3.3.1 Gyro Values Acquisition and Computing the Load on the Robot

For the experiments done there were 2 dual-axis gyros connected to the robot's ATmega Servoboard sensing the direction in which the robot was moving during its walking - by measuring the angular rate of change. All gyros are located in one single place on the robot - above the embedded system, on the S2-HuRo back (Figure 3.13 (e)). The gyro values for the lateral and longitudinal axes were also sent via the serial interface to the PC or to the embedded onboard system for further processing. At each S2-HuRo start, the sensor values were calibrated by reading 400 values, so the null position for the gyro measurements can be found. For each 30 values acquired, a median value is calculated so the light trembling of the robot (due to the interval control of the servos) does not influence the measurements. There was also an option that the measured values were saved on a disk or on the robot's MMC card.

Unfortunately, the load on the servos cannot be measured directly on S2-HuRo. However, tests have shown that a larger load on the robot's legs results in differences between the measured gyro axis value and the real value (probably due to the design of electronic circuit for the analog-digital converters on the ATmega

Servoboard). Namely the gyro value changes always in a positive direction and by simply reading the value, it cannot be distinguished from the normal gyro value. To compensate this, the second gyro is mounted in the opposite direction to the first one. The resulting reading is negative instead of positive for the same lateral or longitudinal movement. It has been observed that similarly to the previous case, an increased load on the robot's servos has an influence on the actual read gyro value, which in this case is drifting in the opposite (negative) direction. From this it can be concluded that the second gyro values can be used in combination with the first gyro values to calculate the load on servos on the originally read gyro values at each moment in time. For calculating such a disturbance in the gyro value, the fact was used, that the difference of two signals with symmetrical differential coding gives the disturbance signal. [PTP03]

In this case we have two gyro signal values "gyro_X_1" and "gyro_X_2" which are read from two gyros for the same axis but positioned oppositely which give the values that include the disturbances in different directions (positive and negative). The following formula gives the disturbance.

$$(\text{gyro\_X\_1} + \text{gyro\_X\_2})/2$$

The gyro values without disturbances are calculated as follows:

$$\text{gyro\_X\_1}_{\text{new}} = \text{gyro\_X\_1} - (\text{gyro\_X\_1} + \text{gyro\_X\_2})/2$$

$$\text{gyro\_Y\_1}_{\text{new}} = \text{gyro\_Y\_1} - (\text{gyro\_Y\_1} + \text{gyro\_Y\_2})/2$$

Similarly, this is also done for the other values.
The process of calculation of disturbance from the gyro values is presented in Figure 6.7.

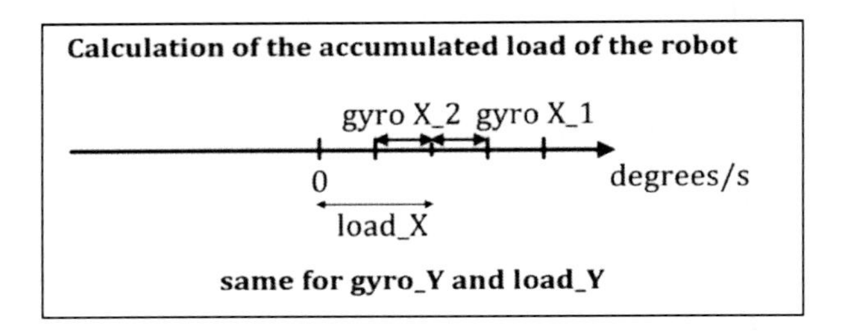

Fig. 6.7 Calculation of the disturbances from the gyro and load values

6.3.3.2 Estimating the Energy Consumption of the Robot's Walking

For optimizing the robot's walking, energy consumption also plays an important factor. The less energy consumed, the better the robot's walking gait because the robot can walk for a longer time. For the S2-HuRo we calculate the energy consumption from the load which can also be computed from the disturbance from the gyro values in either the lateral or longitudinal axis, as previously described:

$$\text{Load\_X} = (\text{gyro\_X\_1} + \text{gyro\_X\_2})/2$$

Or

$$\text{Load\_Y} = (\text{gyro\_Y\_1} + \text{gyro\_Y\_2})/2$$

The load value is calculated in the range from 0 to 70 (Amp x 0.03 units), where the value of 70 units correspond to about 2.1 Amperes of current consumption. The experimentally measured maximum load on the servos by the robot was 2.8 Amp, which divided by 70 unit values gives 0.03. The number of 70 units is chosen in order to be similar to the max values by other sensors measurements (for example by gyros max is by 66 units), which simplifies the computation and data representation.

In a case where the different measured load values are calculated (which is due to random values generated within the read gyro values), this means there is an overload within the servos and in this case the value for the load will be automatically set to the maximum value of 70 units.

The process of calculating the overload is presented in Figure 6.8.

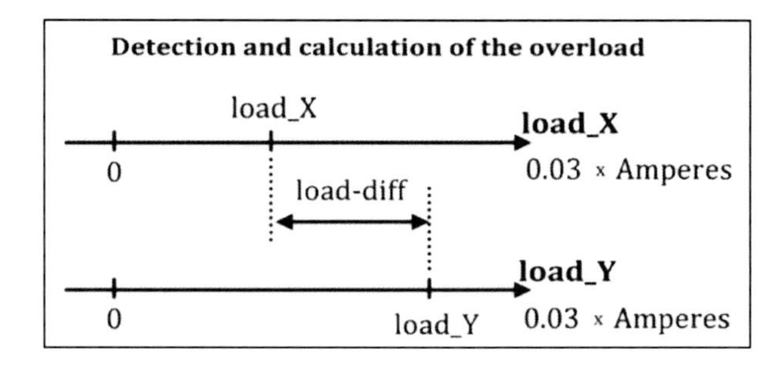

Fig. 6.8 Calculation of overload of the servos.

6.3.3.3 Foot Sensor Contacts and Their Relation to SymbScore Computation

As described in Chapter 3.5 - Figure 3.15, the S2-HuRo robot has 3 binary sensor contacts on each foot. The value of the contact sensors indicates if part or all of the robot's foot is on the ground or not. The binary sensors are directly connected to the robot's ATmega Servoboard and also transferred via serial connection to the PC or to the robot's onboard embedded system.

If the robot falls to the ground when walking, the contacts on the feet indicate that the robot has fallen down and therefore any further movement is disabled to protect the robot's parts from mechanical overloading. The contact sensors also contribute to the *SymbScore* value computation. Namely when the robot falls on the ground indicates that the evaluated set of walking parameters is perhaps not the best choice.

6.3.4 Fuzzy Logic Computation of SymbScore Value

As previously mentioned, fuzzy logic is used by the *SymbScore* value computation.

There are 3 different inputs into the system:

Gyro value for lateral axis;

- Gyro value for longitudinal axis;
- Load within the system. (when there is no overload in the system, the Load_X and Load_Y are the same, so Load_X or Load_Y can refer to the load in the system).

There are 2 intermediate values:

- Stability_X – related to stability computation for one of the robot's axes;
- Stability_Y – related to stability computation for the other of the robot's axes;

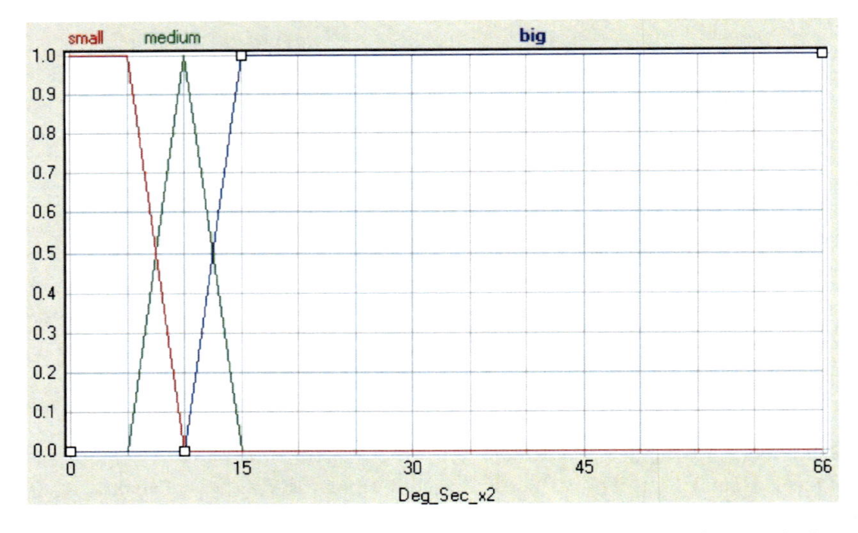

Fig. 6.9 Gyro value for each of the lateral and longitudinal axes mapped in domain from 0 to 66 deg/sec units.

Additionally there are 2 fuzzy logic rule bases:

- "Stability rule base" with fuzzy logic rules for calculating the intermediate Stability_X or Stability_Y values.
- "Symbiosis rules base" with fuzzy logic rules for calculating the output *SymbScore* value.

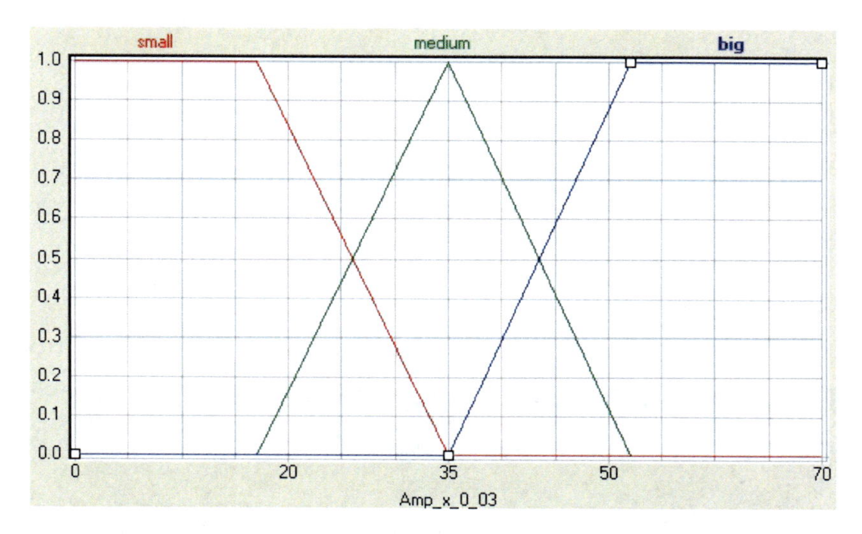

Fig. 6.10 Load value mapped in domain from 70 Amp x 0.03 units.

There are 3 fuzzy logic sets used for each variable.

The read gyro value for each of the lateral and longitudinal axes was mapped in domain from 0 to 66 deg/sec units. The fuzzy sets for gyro value (Figure 6.9) were manually tuned during the pre-experiments done with the robot, which were related to the sensitivity of the gyro and its readings. The load value was mapped in domain of 0 to 70 Amp x 0.03 Units with 3 fuzzy sets evenly distributed (Figure 6.10).

The rule base consists of fuzzy logic rules that have the following premises: gyro value per each of the two axes and load and a consequent part: "Stability_X" value. The fuzzy logic rule base is shown in Figure 6.11. As indicated in the figure, input values are "GYRO_X_LON" and "LOAD" and output value is "STAB_X_LON" and is related to stability value for the longitudinal axis. However, the same rules are also used for computing "Stability_Y" for the robot's lateral axis with inputs "GYRO_Y_LAT" and "LOAD".

#	IF GYRO_X_LON	LOAD	THEN DoS	STAB_X_LON
1	small	small	1.00	stable
2	medium	small	1.00	medstable
3	big	small	1.00	nonstable
4	small	medium	1.00	medstable
5	medium	medium	1.00	medstable
6	big	medium	1.00	nonstable
7	small	big	1.00	medstable
8	medium	big	1.00	medstable
9	big	big	1.00	nonstable

Fig. 6.11 Rule base of fuzzy logic rules that is associated for computing of the Stability value for each of the lateral and longitudinal axes.

The "non-stable" situations are indicated when the gyro values "big", which relates to situations when the robot falls on the ground.

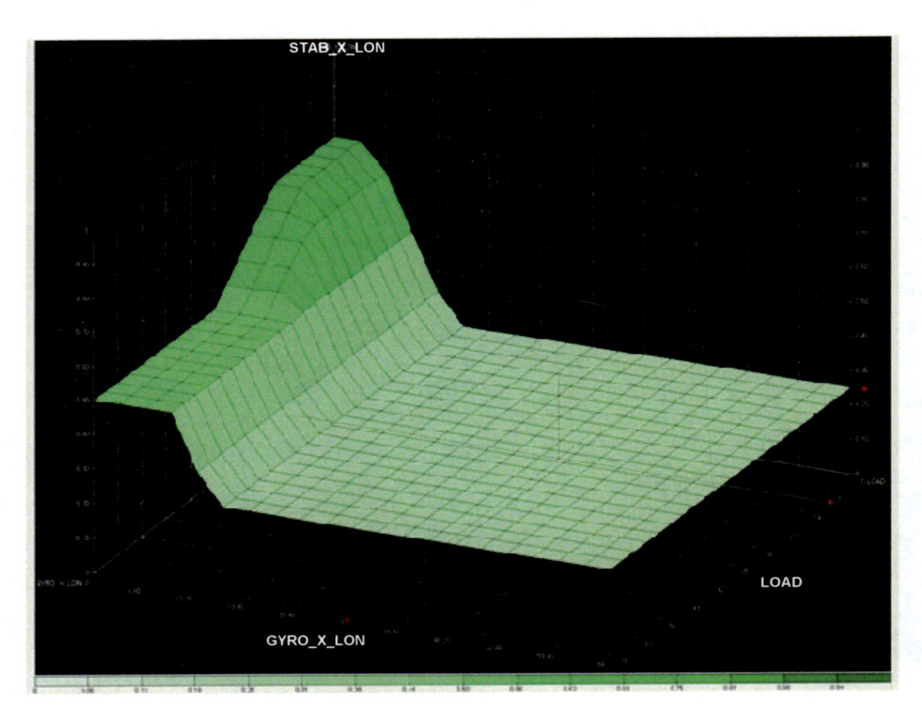

Fig. 6.12 3D representation of the computational surface for "Stability_X".

There is a 3D representation of the computational surface for “Stability_X” presented in Figure 6.3.4.4. The lower plane x and y axes are “GYRO_X_LON” and “LOAD” - related to gyro value and load value for the longitudinal axis. The z axis is the “STAB_X_LON” representing the surface for computing the “Stability_X” value output, with respect to the input gyro and load values. This is also similar to the surface used for computation of the “Stability_Y”.

The fuzzy sets of the “Stability_X” and “Stability_Y” (Figure 6.3.4.5) are distributed evenly in domain from 0 to 1 Stability Units. The “Stability Units” don’t have real world interpretation, and are only used as intermediate values for computation of the final *SymbScore* value.

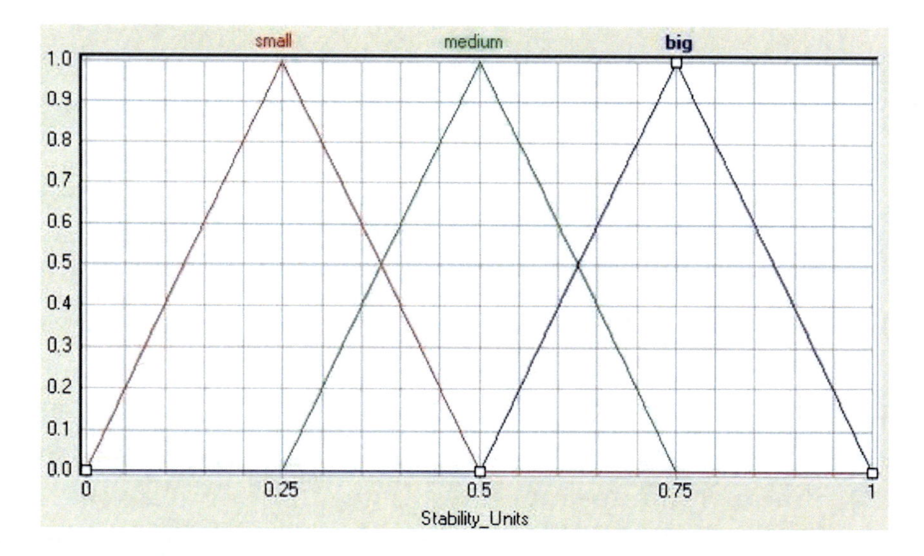

Fig. 6.13 The intermediate variables “Stability_X” and “Stability_Y” each have fuzzy sets and mapping from 0 to 1 units.

The *SymbScore* computed value is mapped in the domain from 0 to 1, where a value of 0 is stable robot walking and a value of 1 indicates non-stable robot walking. It has 3 fuzzy logic sets: nonstable, medstable and stable (Figure 6.14).

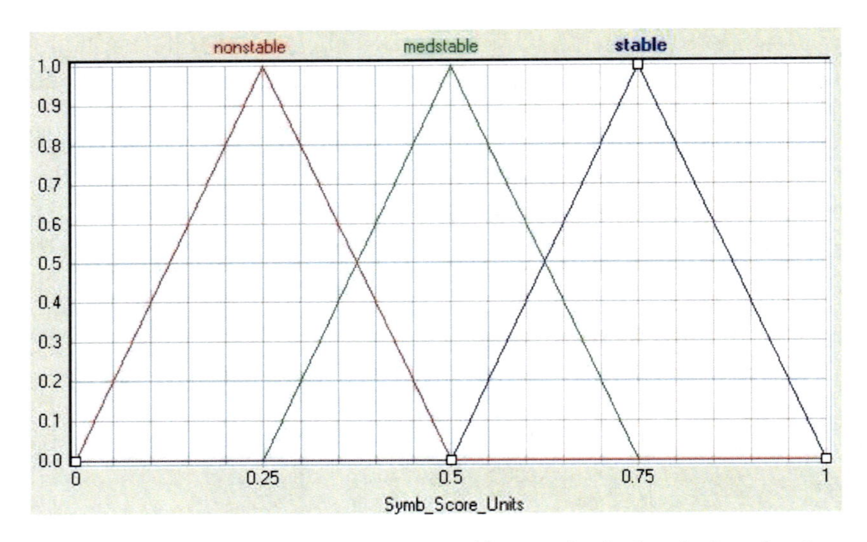

Fig. 6.14 *SymbScore* computed value and its mapping in domain from 0 to 1.

The main part of computation of the *SymbScore* value is related to the Symbiosis inspiration. The fuzzy logic rules involved for computation of *SymbScore* are shown in Figure 6.3.4.7. In the figure there are 2 highlighted rules which directly explain the "symbiosis relation" between the longitudinal and lateral axes.

#	IF STAB_X	STAB_Y	THEN DoS	SYMB_SCORE
1	small	small	1.00	small
2	medium	small	1.00	medium
3	big	small	1.00	small
4	small	medium	1.00	medium
5	medium	medium	1.00	medium
6	big	medium	1.00	big
7	small	big	1.00	small
8	medium	big	1.00	big
9	big	big	1.00	big

Fig. 6.15 Rule base of fuzzy logic rules for computing the SymbScore value inspired by symbiosis.

The firstly highlighted symbiosis rule is defined as:

- IF "STAB_X" is "big" and "STAB_Y" is "small" then "SYMB_SCORE" is small.

This rule can be interpreted as when the stability in the longitudinal axis is rather big, indicating stable robot movement in longitudinal axis. However, the stability in the lateral axis is rather small, indicating unstable robot movement in lateral axis. As a result, the computed *SymbScore* is rather small, meaning the robot is rather unstable although the robot's movement in one axis is stable.

The second highlighted symbiosis rule is defined as:

- IF "STAB_X" is "small" and "STAB_Y" is "big" then "SYMB_SCORE" is small.

This rule can be interpreted as when the stability in the longitudinal axis is rather small, indicating unstable robot movement in lateral axis. However, the stability in the longitudinal axis is rather big, indicating stable robot movement in the longitudinal axis. As a result, the computed *SymbScore* is rather small, meaning the robot is rather unstable although the robot's movement in one axis is stable.

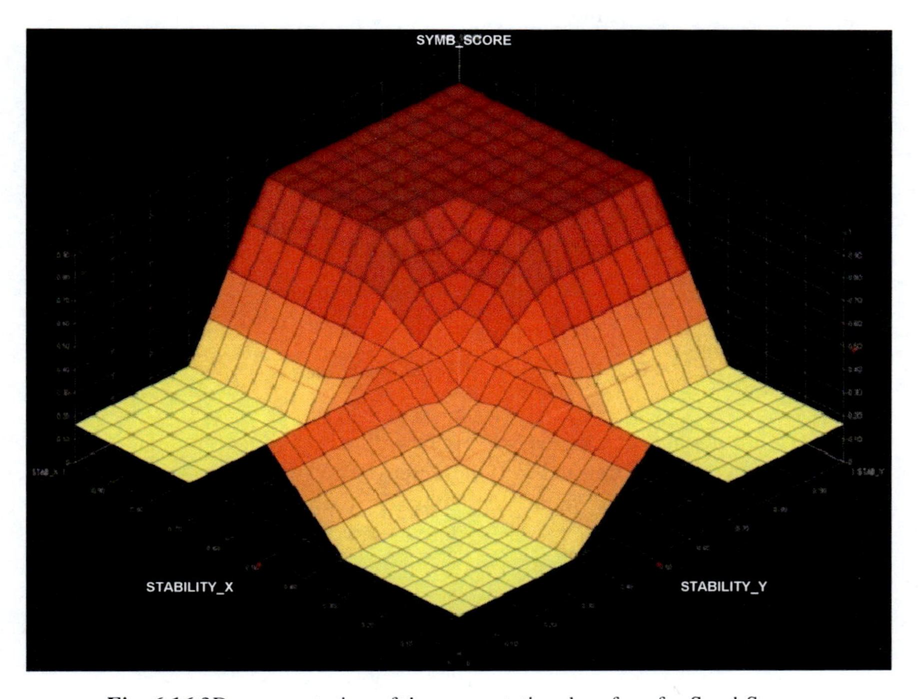

Fig. 6.16 3D representation of the computational surface for SymbScore.

The 3D computational surface for *SymbScore* is presented in Figure 6.16. The x, y plane axes are "STAB_X" and "STAB_Y" - related to the intermediate computed values for stability in lateral and longitudinal axes. The z axis is "SYMB_SCORE" representing the *SymbScore* value with respect to the input intermediate stability values.

6.3.5 Genetic Algorithm Details for the SelSta Approach

The genetic algorithm is used to select the next generation of improved walking parameters (lateral and longitudinal positioning of the robot foot) for balancing the movement of the robot.

The score - *SymbScore* is used as a fitness function result for the genetic algorithm.

In each self-stabilizing robot run, there may be several such evaluation periods up to the moment where enough optimized self-stabilizing walking of the robot is generated for a particular surface.

The parameters for the genetic algorithm are presented in Figure 6.17. Each individual in the population of the genetic algorithm has the following format:

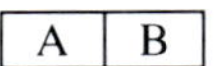

Where A, B represent: lateral (A) and longitudinal (B) leg positioning of the normal robot still in standing position and range from -5 to 5 degrees, with a resolution of 0.5.

Genetic Algorithm Parameters	
number of generations	15
population size	10
replacement percentage	0.5
convergence percentage	0.99
crossover probability	0.6
mutation probability	0.05

Fig. 6.17 Genetic algorithm parameters.

The resolution can be decreased if needed. The A, B parameters for the standing still position are defined as 0, 0 degrees. The genetic algorithm is a single point crossover. The replacement percentage is 0.5 - meaning one half of the population in every cycle is replaced with a new one. The population size is initialized at the start to 10 genomes each holding the A, B parameters for lateral and longitudinal foot position.

The genetic algorithm is run only whenever the *SymbScore* value is first computed.

The genetic algorithm finishes with its search either when the number of generations reaches 15 or the convergence percentage is 0.99. At the end, optimized parameters are found for balancing the movement of the robot walking on the particular surface.

6.3.6 Preparation for Experiments

In each experiment the robot optimizes its walking over a number of test sections. Each test section consists of 6 walking cycles, of which the first and last walking ones are related to starting and stopping, respectively. The walking cycle can be

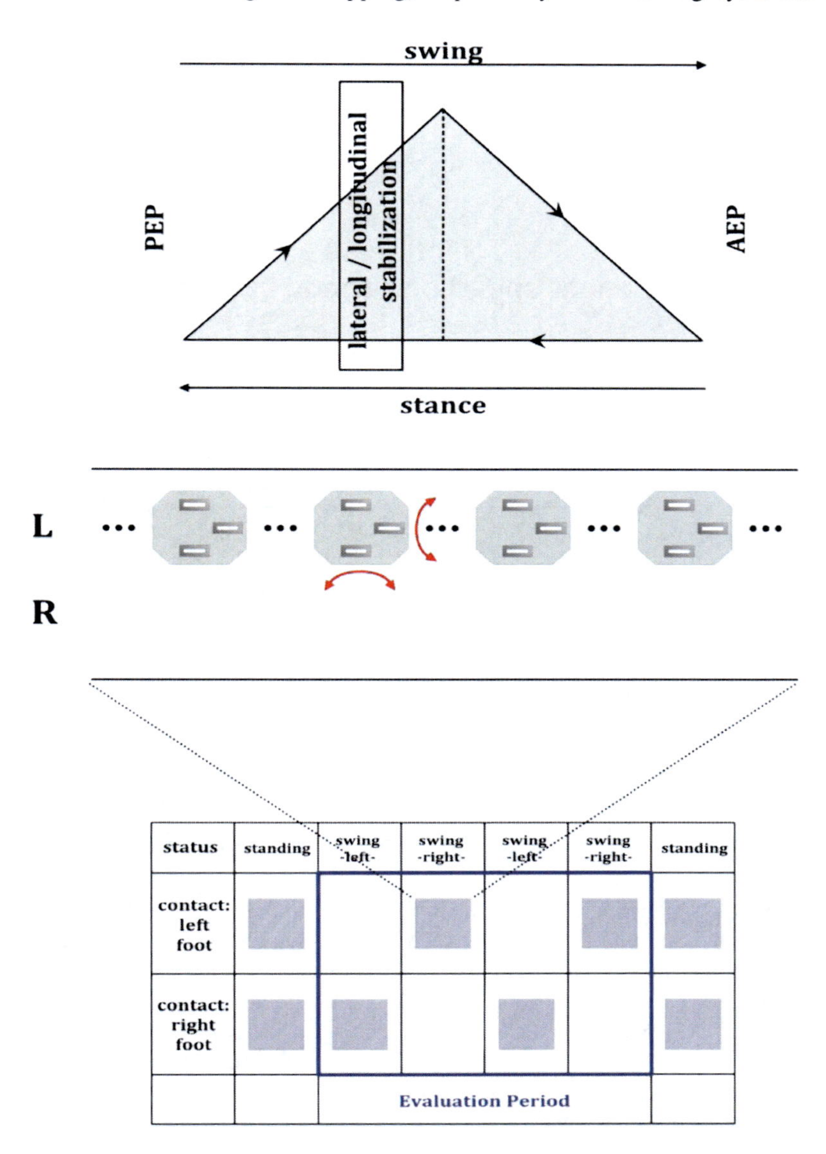

Fig. 6.18 Graphical representation of test section with corresponding walking cycles. L and R indicate for left and right legs respectively.

described as a situation in which one leg is in its stance phase and the other leg is in its swing phase. Before each walking section, the robot sets the parameters for lateral and longitudinal balancing movement that takes place in about the middle of every walking cycle (depicted with round lines with arrows in Figure 6.18).

The beginning and ending cycles are not included in evaluation of the *SymbScore* value, only the middle 4 walking cycles are considered for computation. The *SymbScore* value is computed at the end of the test section as an average of the 4 *SymbScore* values computed during each of the 4 walking cycles for that particular test section. The SymbScore value of 0 represents that the robot has fallen during that test section.

One such test section is represented in Figure 6.19.

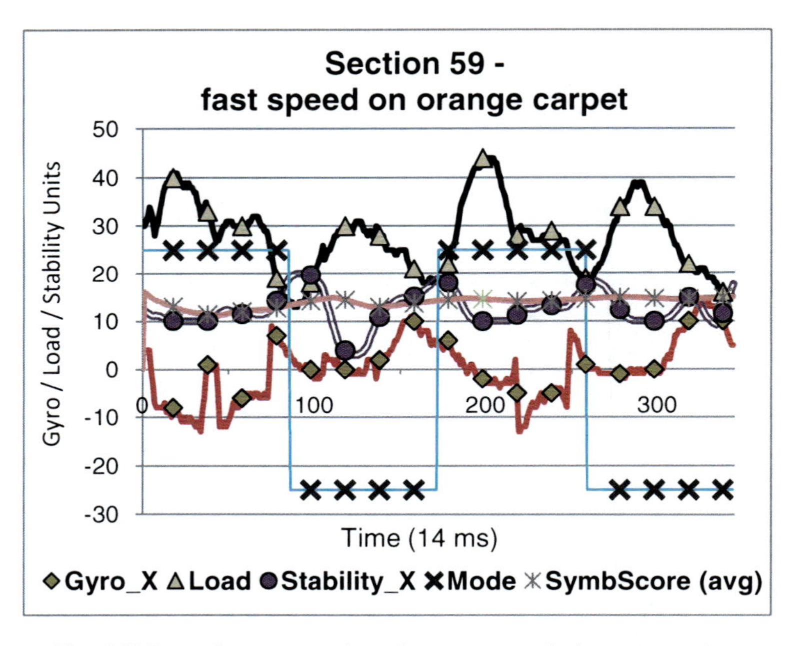

Fig. 6.19 Example representation of measurement during a test section.

For better visualization, the values in this figure and all other related experiment figures, map "Stability_X", "Stability_Y" and "SymbScore" (averaged) 0 to 1 range domain values into 0 to 20 range domain. The values of "Gyro_X", "Gyro_Y" and "Load" are not resized and are shown as measured during the experiment. The "Mode" value shows which leg is in its swing phase (while the other is in its stance phase). The value of 25 means that the left foot is in its swing phase and the right foot in its stance phase, while the value of -25 means the right foot is in its swing phase and the left foot is in its stance phase. "Mode" has value of 0 when the robot has fallen.

It is also important to know that in the experiments there are 3 threads running on the PC of which one named "sensor thread." The sensor thread is related to the sensor values acquisition and delivers new data in intervals from 55-65ms and is also responsible for data logging.

The acquired sensor data is delivered to the "Movement strategy thread" which computes the *SymbScore* value during the stability and load robot evaluation during the test section. At the end of the test section it computes an average *SymbScore* value for that test section and forwards this to the "Genetic algorithm thread" which then waits for the *SymbScore* value.

After receiving the *SymbScore* value, the "Genetic algorithm thread" forwards the next gene values related to lateral and longitudinal stabilizing movement of the robot, which need to be evaluated in the next test walking section.

The "Movement strategy thread" also waits until the "Genetic algorithm thread" is finished with the genome evaluation. The waiting time is very short due to the low population, and this means that such threads, including the genetic algorithm, can also be run on an embedded system onboard the robot, so the self-stabilization of the robot can be evaluated under real circumstances and on a real robot.

This is presented in Figure 6.20.

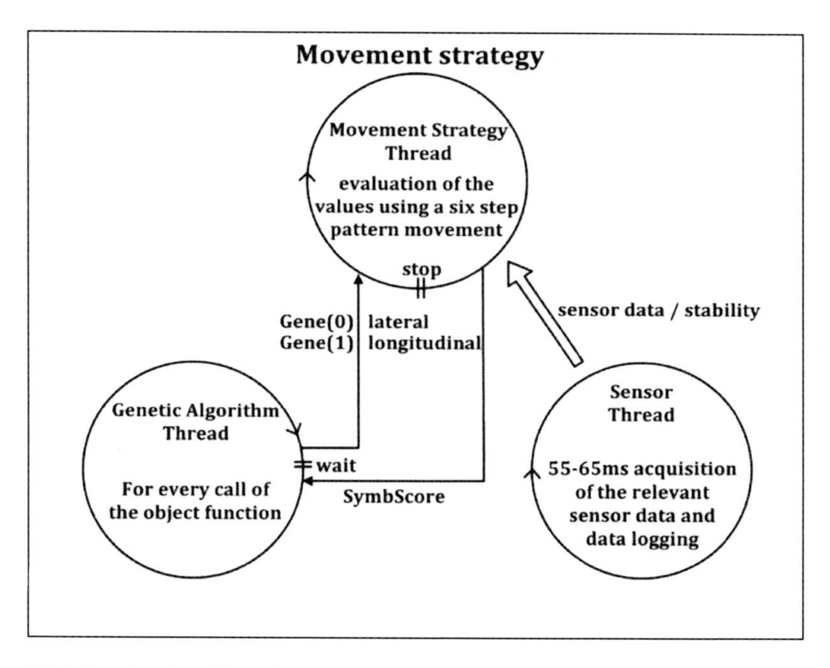

Fig. 6.20 Genetic algorithm, the movement strategy, sensor acquisition, and SymbScore generation.

6.4 Experiments Done with the SelSta Approach

In order to conduct as many experiments as possible and to test the usefulness of the SelSta approach, all the experiments presented here are done on a PC computer (instead of on the Gumstix® embedded system, since the data logging from the experiments is faster on the PC than on the embedded system) which was connected to the ATmega servo controller running a program that delivers the sensor data to the PC via a serial cable. The PC was also used to start and stop the robot's walking.

For testing the self-stabilizing approach on S2-HuRo, several tests were made on four different flat surfaces – 3 carpets with different thickness and softness levels and also one linoleum type of hard surface. Different surfaces were tested while performing self-stabilized walking, since the surfaces introduce dynamics to the robot's walking due to their own material properties.

Each time the optimal parameters were found with the SelSta approach (using the genetic algorithm) for a particular surface/carpet, an additional test was performed on those optimally found walking parameters in 5 walking test sections and the best result was selected for later comparison.

On each surface, the SelSta approach tried to find the optimal walking parameters for 3 different speeds: slow, medium, and fast speed as mentioned in Section 6.3.2.

For comparing the results and the usefulness of the SelSta approach, the results from the SelSta based tests were compared with results from tests made with "Standard walking" parameters and "Manually optimized" walking parameters as described in Section 6.3.2.

The robot is hung on a steel cable via metal rings during the experiments. The rings give the robot enough space for performing its walking actions without influencing the walking movement itself. On the other hand they give support when the robot falls due to some improper walking behavior or poorly generated balancing parameters. When this happens, a human operator puts the robot in the standard standing position first. Then via PC command the robot is instructed to continue with the new cycle of balancing movement parameter generation with the *SelSta* approach until the optimal parameters for balancing movement are found. This approach with the steel wire rope was chosen since it was guessed that the robot would fall often during the experiments, however the experiments later proved that this was overcautious.

In case the robot falls during its walking and is on the ground for about 20 foot sensor readings (indicating a 0 value or no contact with the ground) the robot's position is reset to standing position and the human operator brings the robot to its feet again. In a real case scenario the help from a human operator can be avoided by enabling the robot to get into its standing position by itself and get oriented in space and then continue with the self-stabilized walking. However, for the rather large number of experiments it was decided that the robot's stand up after a fall should be assisted by the human operator.

If the robot has just started its walking step when given a "stop command", then the leg movement proceeds until the middle point of the walking step is done, and then stops. If the "stop command" is issued when the leg movement is in its second half of the walking step, then the leg which is in its swing phase returns back to the middle point of its swing trajectory, and when the leg reaches the ground the robot is brought to its still stand position with both feet on the ground. The following is a detailed description of each of the tests done with the self-stabilized walking approach SelSta. These tests with the S2-HuRo robot were performed on several flat surfaces:

- soft carpet;
- medium soft carpet;
- hard carpet;
- hard linoleum surface.

At the end of this chapter, a summary and comparison is given of the final results from the experiments done.

6.4.1 Experiments on a Soft Green Carpet

Experiments were done on a soft green carpet (Figure 6.21) to test how the SelSta approach would be able to find the optimal self-stabilizing walking on such a carpet which due to its thickness introduces rather large dynamics in the robot's walking. Figure 6.22 shows the test setup with the S2-Huro robot during the experiment on the soft green carpet.

Fig. 6.21 Soft green carpet.

Fig. 6.22 Robot S2-HuRo on a soft green carpet.

6.4.1.1 Results from SelSta Optimal Walking Parameters Search on a Soft Green Carpet

The results from the experiments done with SelSta on a soft green carpet are divided into 3 different categories - depending on the speed of the robot during the experiments: "Slow speed", "Medium speed" and "Fast speed".

Slow speed

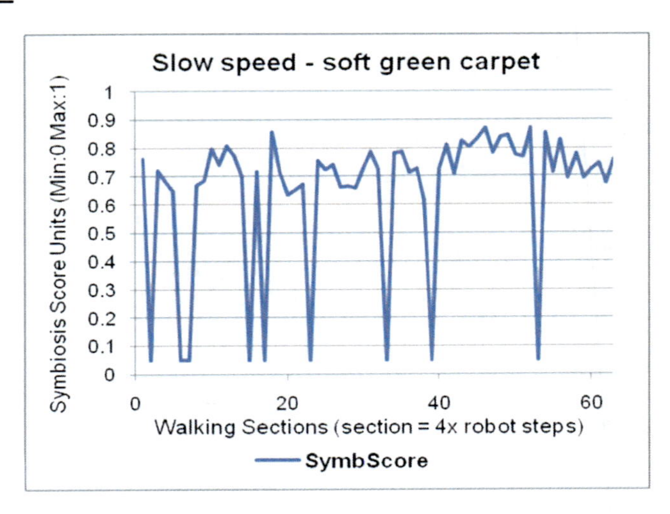

Fig. 6.23 SymbScore of slow walking speed of S2-HuRo over a soft green carpet.

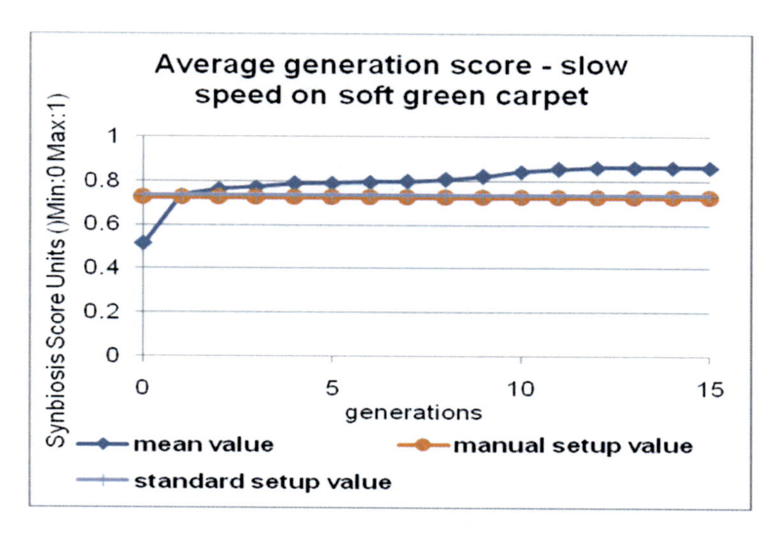

Fig. 6.24 Average SymbScore during genetic generations for slow walking speed of S2-HuRo over a soft green carpet.

Optimal walking parameters (for impulsive stabilizing movement) for slow walking speed of the S2-HuRo on a soft green carpet were found within 63 walking sections done in 16 min. The best walking stabilization parameters found were 1.5 degrees for longitudinal and 0.5 degrees for lateral impulsive feet movement.

Medium speed

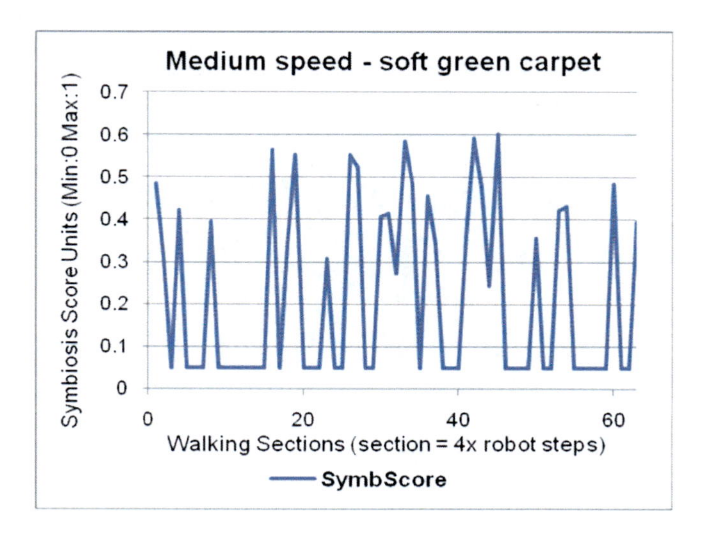

Fig. 6.25 SymbScore for medium walking speed of S2-HuRo over a soft green carpet.

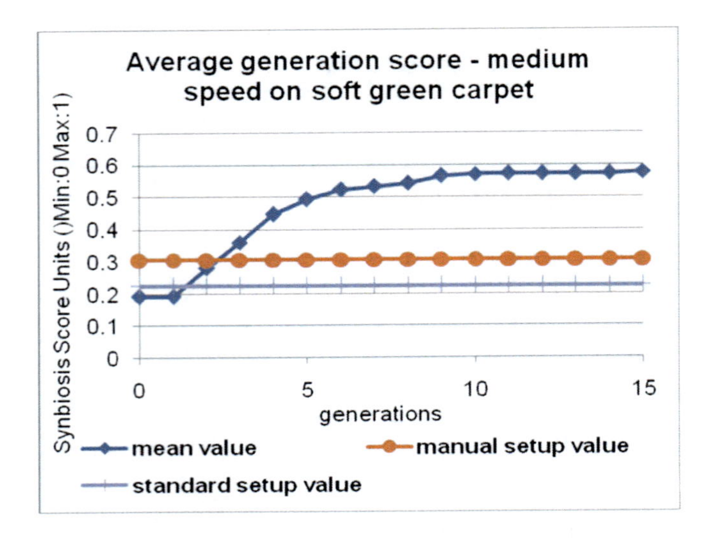

Fig. 6.26 Average SymbScore during genetic generations by medium walking speed of S2-HuRo over a soft green carpet.

Optimal walking parameters (for impulsive stabilizing movement) for medium walking speed of the S2-HuRo on a soft green carpet were found within 69 walking sections completed in 14 min. The best walking stabilization parameters found were 2 degrees for longitudinal and 4 degrees for lateral impulsive feet movement.

Fast speed

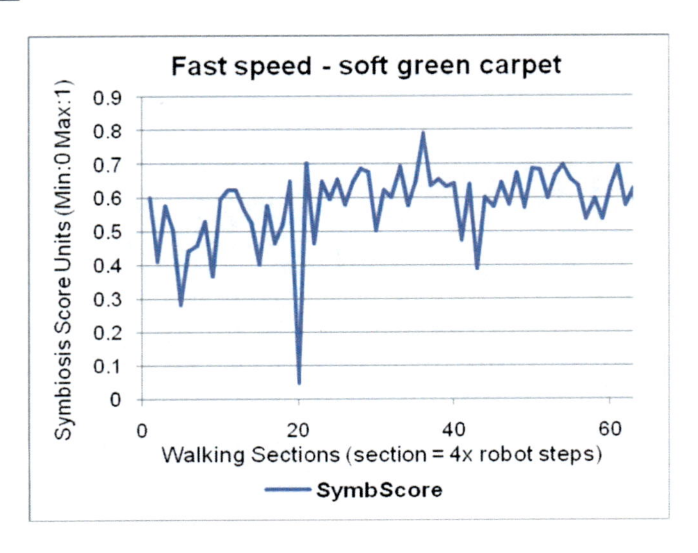

Fig. 6.27 SymbScore for fast walking speed of S2-HuRo over a soft green carpet.

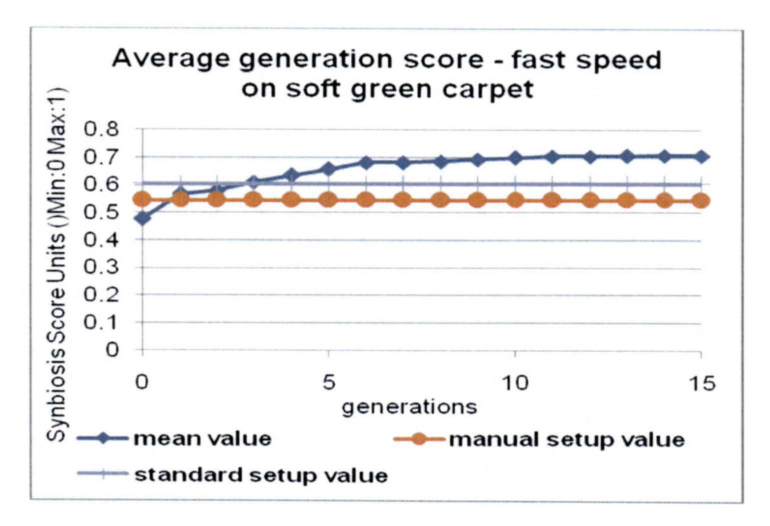

Fig. 6.28 Average SymbScore during genetic generations for fast walking speed of S2-HuRo over a soft green carpet.

Optimal walking parameters (for impulsive stabilizing movement) for fast walking speed of the S2-HuRo on a soft carpet were found within 67 walking sections completed in 12 min. The best walking stabilization parameters found were 0.5 degrees for longitudinal and -4.5 degrees for lateral impulsive feet movement.

6.4.1.2 Discussion about the Walking Sections Results and SymbScore Values throughout Genetic Generations by Self-stabilization on a Soft Green Carpet

From the walking sections of "Slow speed", "Medium speed" and "Fast speed" (Figure 6.23, Figure 6.25, Figure 6.27) it can be seen that by using the SelSta approach the robot's fallings (the lowest *SymbScore* values on the graphs) are reduced during the experiment.

Averaged *SymbScore* (Figure 6.24, Figure 6.26, Figure 6.28) measured over generations and the direct comparison is made with the best *SymbScore* measured values found during the 5 tests done with "Standard walking" parameters and "Manually optimized walking" parameters. After about 4 generations, the *SymbScore* value is higher than the best values found in "Standard walking" and in walking with "Manually optimized" walking parameters. After the 12th generation it can be seen that the *SymbScore* value in the experiments with SelSta approach does not continue to increase.

6.4.1.3 Comparison of the Best Found Values by "SelSta Walking" with the "Standard Walking" and "Manually Optimized" Walking by Slow, Medium and Fast Walking Speeds on a Soft Green Carpet

Once the *SelSta* approach finds the best parameters for optimal walking with respect to the robot's stability and energy consumption during walking, 5 additional walking tests are made with those parameters and the measurements from best walking test (with the highest average *SymbScore*) are compared with the best walking test results from the 5 tests conducted with "Standard walking" (non calibrated walking parameters) and the 5 tests conducted with "Manually optimized walking" (manually calibrated walking parameters). Only the best of the "Standard walking"/"Manually optimized walking" test values were selected to be compared with results from the *SelSta* approach.

As can be compared from the results in figures Figure 6.29, Figure 6.30 the "Load" peak values of walking parameters found with the *SelSta* approach are smaller during the robot's walking than the "Load" peak values when walking with "Standard walking" parameters. Also the "Gyro_X" average values from the *SelSta* approach are smaller than the "Gyro_X" average values from the "Standard walking" parameters (The capped-off gyro values show overloading by the servos at those particular moments in time). All this indicates that the walking parameters found with the *SelSta* approach for slow speed on a soft green carpet can introduce more stable and energy efficient walking than with non-calibrated "Standard walking" parameters.

Slow speed

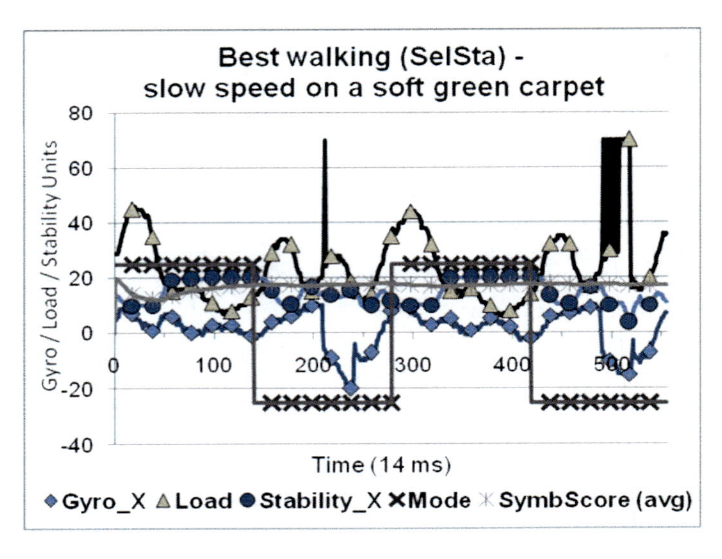

Fig. 6.29 Best walking section with SelSta approach by slow robot walking speed on a soft green carpet.

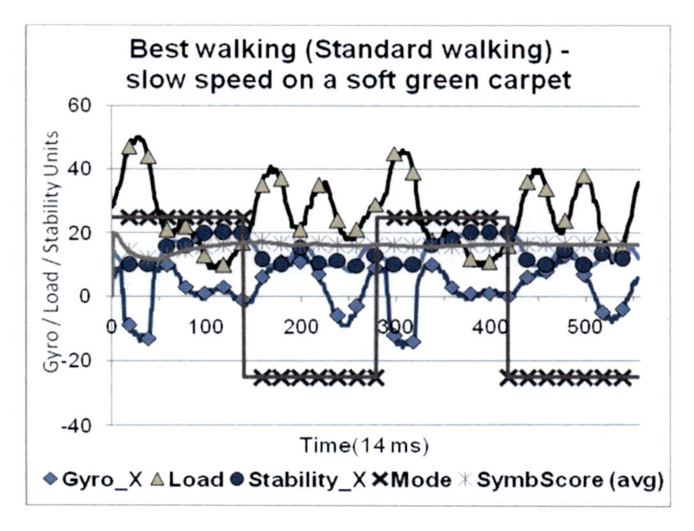

Fig. 6.30 Best walking section with "Standard walking" parameters for slow robot walking speed on a soft green carpet.

From the results in figures Figure 6.31, Figure 6.32, the "Load" peak values for walking parameters found with the *SelSta* approach are smaller during the robot's walking than the "Load" peak values for walking with "Manually optimized" parameters (The capped-off gyro values signify a overloading of the servos at those particular moments in time).

Medium speed

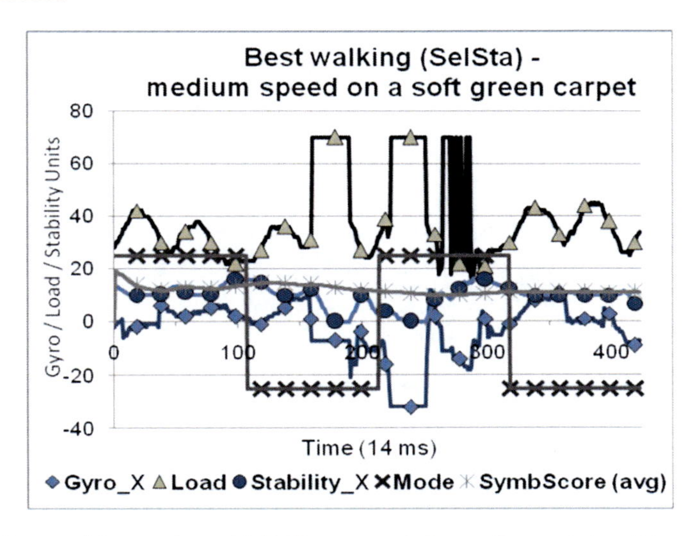

Fig. 6.31 Best walking section with SelSta approach for medium robot walking speed on a soft green carpet.

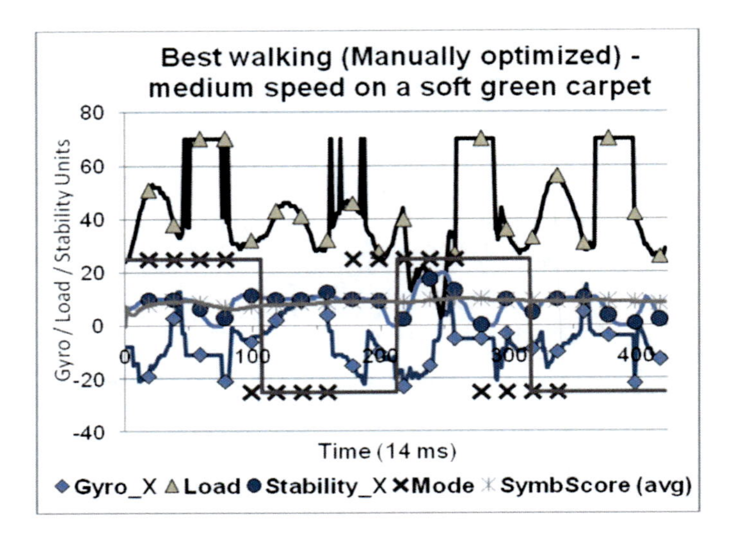

Fig. 6.32 Best walking section with "Manually optimized" walking parameters by medium robot walking speed on a soft green carpet.

Also, the "Gyro_X" average values from the *SelSta* approach are smaller than the "Gyro_X" average values from walking with "Manually optimized" parameters. All this indicates that the walking parameters found with the *SelSta* approach at medium speed on a soft green carpet can introduce more stable and energy efficient walking than by walking with "Manually optimized" walking parameters.

Fast speed

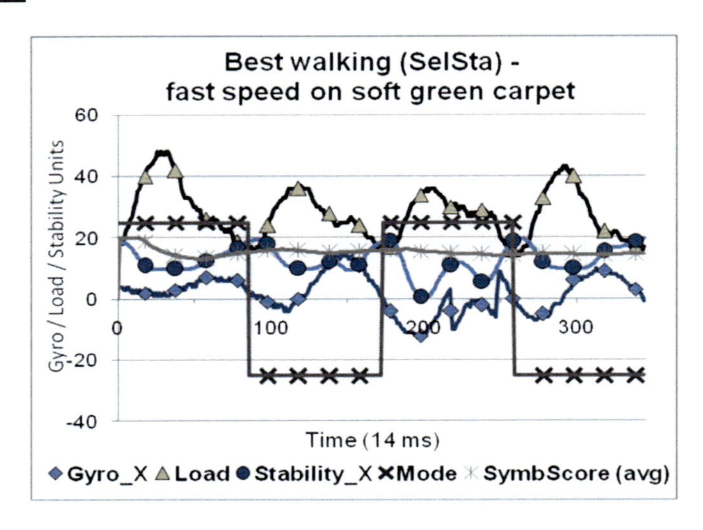

Fig. 6.33 Best walking section with SelSta approach for fast robot walking speed on a soft green carpet.

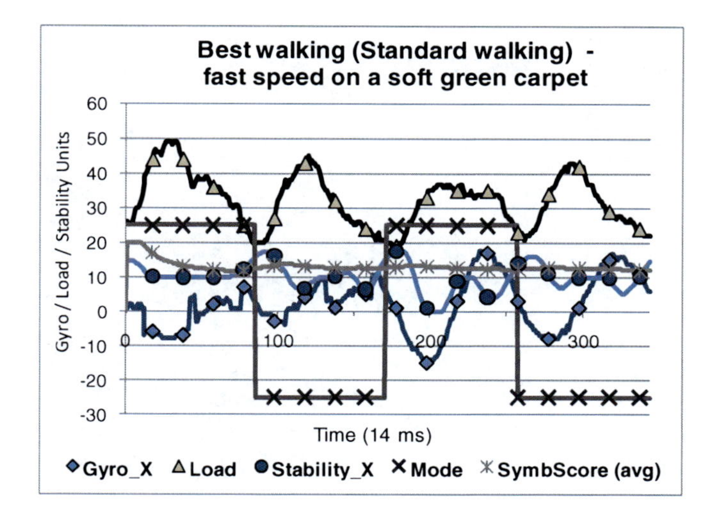

Fig. 6.34 Best walking section with "Standard walking" parameters for fast robot walking speed over a soft green carpet.

From the results in Figure 6.33 and Figure 6.34, the "Load" peak values from walking parameters found with the *SelSta* approach are more or less similar to the "Load" peak values from walking with "Standard walking" parameters. Also, the "Gyro_X" average values from the *SelSta* approach are more or less similar to the "Gyro_X" average values for walking with "Standard walking" parameters. All this indicates that the walking parameters found with the *SelSta* approach for fast speed on a soft green carpet can in most of the cases introduce more stable and energy efficient walking than the walking with "Standard walking" parameters.

6.4.2 Experiments on a Medium Soft Orange Carpet

Experiments were done on a medium soft orange carpet (Figure 6.35) to test out how well the SelSta approach will be able to find the optimal self-stabilizing walking on such a carpet which due to its medium thickness introduces some dynamics in the robot's walking. Figure 6.36 shows the test setup with the S2-Huro robot during the experiment on a medium soft orange carpet.

Fig. 6.35 Medium soft orange carpet.

Fig. 6.36 Robot S2-HuRo on a medium soft orange carpet.

6.4.2.1 Results from SelSta Optimal Walking Parameters Search on a Medium Soft Orange Carpet

The results from the experiments done with SelSta on a medium soft orange carpet are divided into 3 different categories depending on the speed of the robot during the experiments: "Slow speed", "Medium speed" and "Fast speed".

Slow speed

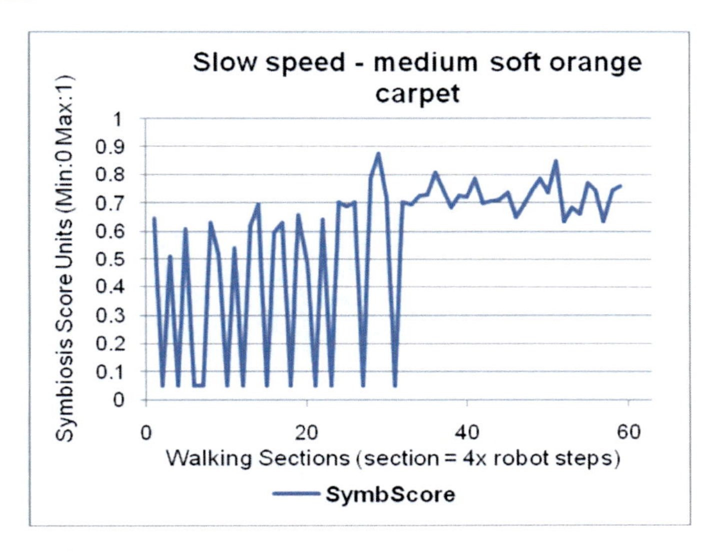

Fig. 6.37 SymbScore for slow walking speed of S2-HuRo on a medium soft orange carpet.

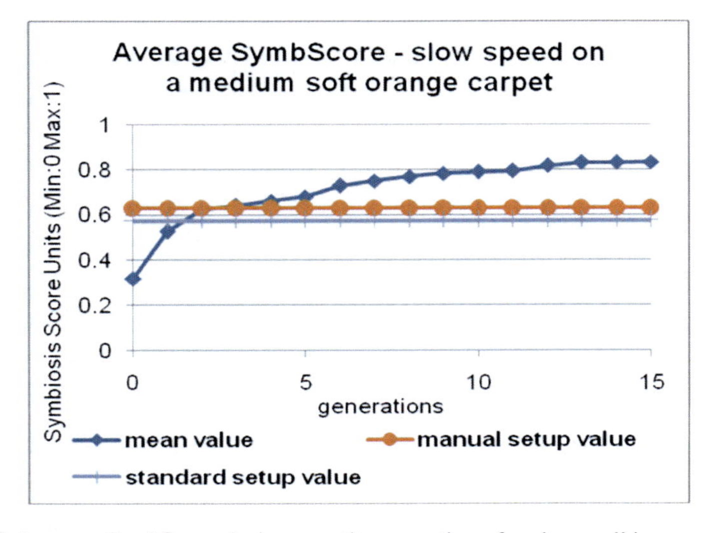

Fig. 6.38 Average SymbScore during genetic generations for slow walking speed of S2-HuRo on a medium soft orange carpet.

Optimal walking parameters (for impulsive stabilizing movement) for slow walking speed of the S2-HuRo on a medium soft carpet were found within 65 walking sections completed in 16 minutes. The best walking stabilization parameters found were 1.5 degrees for longitudinal and 4 degrees for lateral impulsive foot movement.

Medium speed

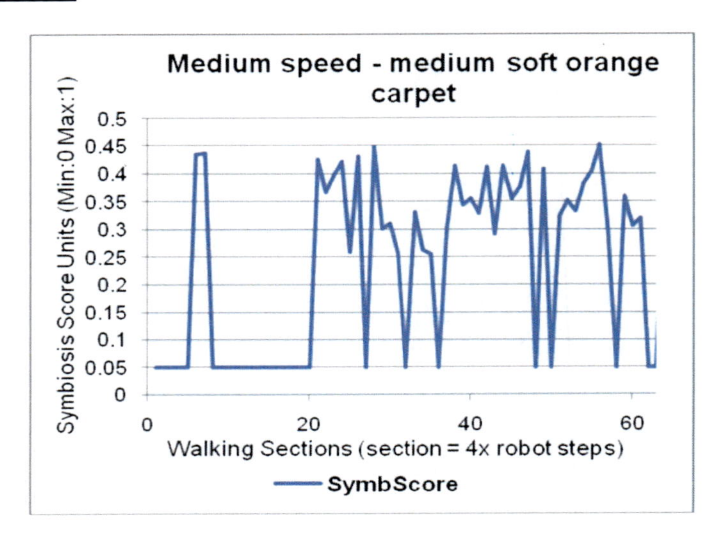

Fig. 6.39 SymbScore for medium walking speed of S2-HuRo on a medium soft orange carpet.

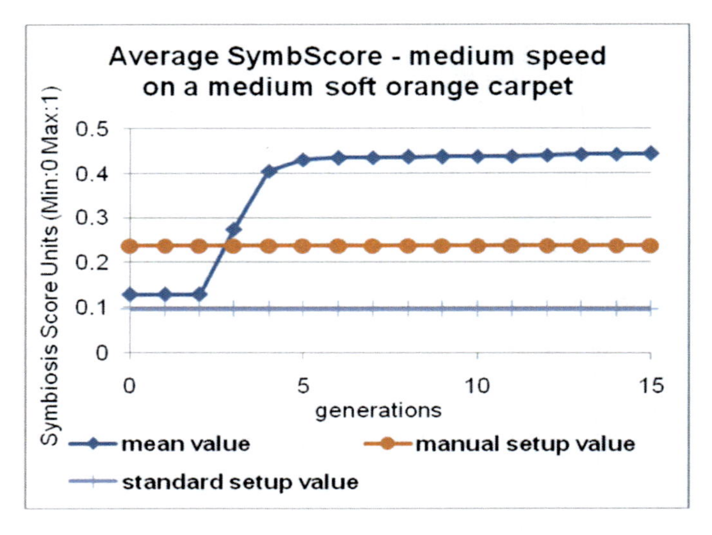

Fig. 6.40 Average SymbScore during genetic generations for medium walking speed of S2-HuRo on a medium soft orange carpet.

Optimal walking parameters (for impulsive stabilizing movement) for medium walking speed of S2-HuRo on a medium soft carpet were found within 65 walking sections completed in 14 minutes. The best walking stabilization parameters found were 1.5 degrees for longitudinal and 4 degrees for lateral impulsive foot movement.

Fast speed

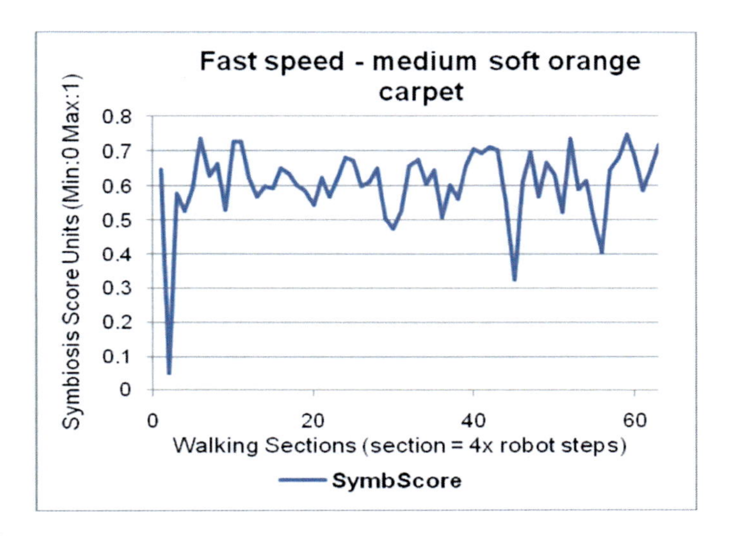

Fig. 6.41 SymbScore for fast walking speed of S2-HuRo on a medium soft orange carpet.

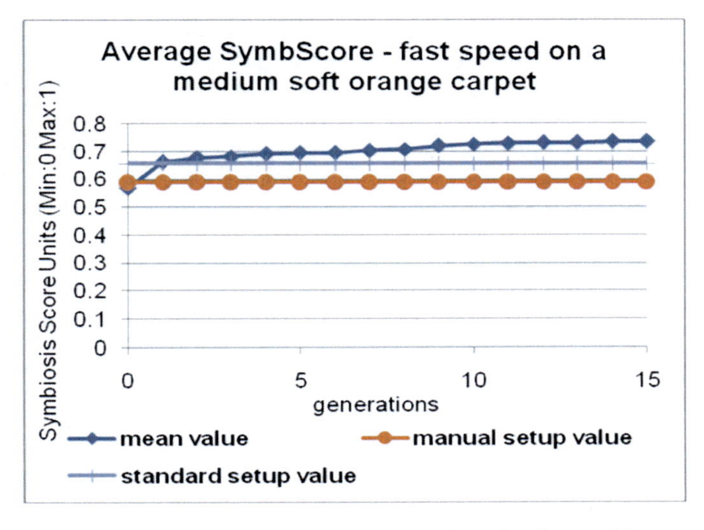

Fig. 6.42 Average SymbScore during genetic generations for fast walking speed of S2-HuRo on a medium soft orange carpet.

Optimal walking parameters (for impulsive stabilizing movement) for fast walking speed of S2-HuRo on a medium soft carpet were found within 66 walking sections completed in 11 minutes. The best walking stabilization parameters found were 1.5 degrees for longitudinal and 0 degrees for lateral impulsive foot movement.

6.4.2.2 Discussion about the Walking Sections Results and SymbScore Values throughout Genetic Generations by Self-stabilization on a Medium Soft Orange Carpet

From the walking sections for "Slow speed", "Medium speed" and "Fast speed" (Figure 6.39, Figure 6.41, Figure 6.43) it can be seen that when using the SelSta approach the robot fallings (the lowest *SymbScore* values on the graphs) are reduced during the experiment.

Averaged *SymbScore* (Figure 6.38, Figure 6.40, Figure 6.42) measured over generations and direct comparisons are made with the best *SymbScore* measured values found by the 5 tests done with "Standard walking" parameters and "Manually optimized walking" parameters. After about 4 generations, the *SymbScore* value is higher than the best values found in "Standard walking" and walking with "Manually optimized" walking parameters. After the 12th generation the *SymbScore* values from the experiments with the SelSta approach do not increase further.

6.4.2.3 Comparison of the Best Found Values by "SelSta Walking" with the "Standard Walking", "Manually Optimized" Walking by Slow, Medium and Fast Walking Speeds on a Medium Soft Orange Carpet

Once the *SelSta* approach finds out the best parameters for optimal walking with respect to robot's stability and energy consumption during walking, 5 additional walking tests are made with those parameters and the measurements from best walking test (with highest average *SymbScore*) have been compared with the best walking test results from the 5 tests conducted with "Standard walking" (non calibrated walking parameters) and 5 tests conducted with "Manually optimized walking" (manually calibrated walking parameters). Only the best of "Standard walking"/"Manually optimized walking" having the best walking test value has been selected to be compared with results from the *SelSta* approach.

As can be compared from the results in figures Figure 6.43, Figure 6.44 the "Load" peak values by walking parameters found with the *SelSta* approach are relatively similar with the "Load" peak values by walking with "Manually optimized" walking parameters (The capped-off gyro values signalize for a overloading by servos at those particular moments of time). Also the "Gyro_X" average values by *SelSta* approach are rather smaller than the "Gyro_X" average values by walking with "Manually optimized" walking parameters. All this indicates that the walking parameters found with the *SelSta* approach by slow speed on a medium soft orange carpet can introduce more stable and energy efficient walking than by walking with "Manually optimized" parameters.

Slow speed

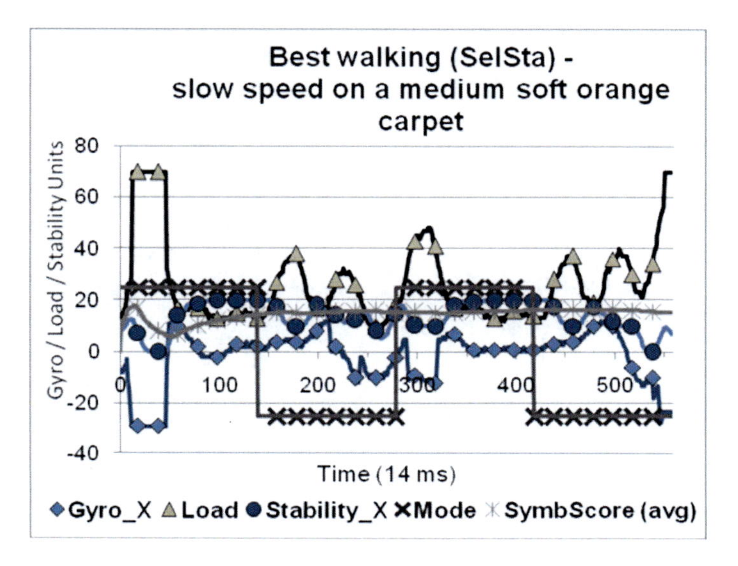

Fig. 6.43 Best walking section with SelSta approach by slow robot walking speed on a medium soft orange carpet.

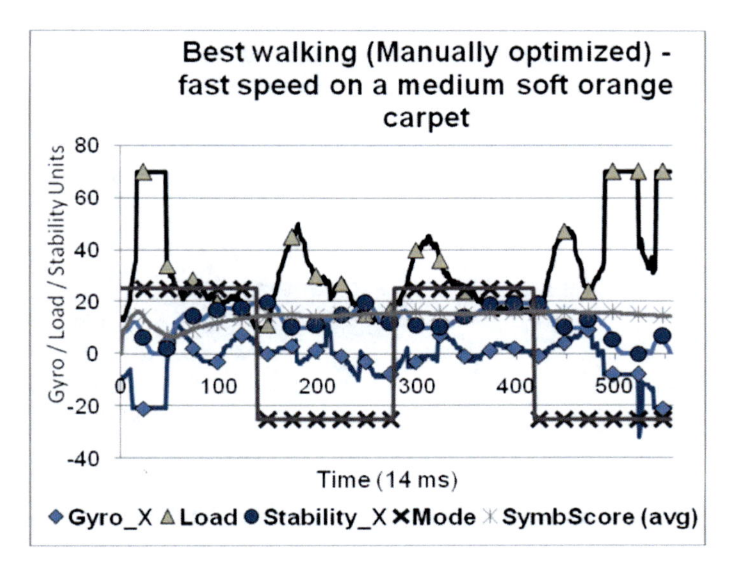

Fig. 6.44 Best walking section with "Manually optimized" walking parameters by slow robot walking speed on a medium soft orange carpet.

Medium speed

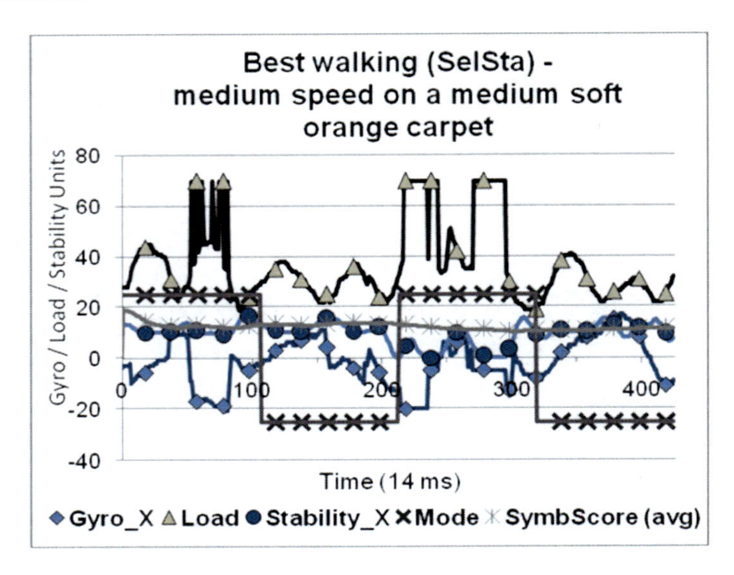

Fig. 6.45 Best walking section with SelSta approach by medium robot walking speed on a medium soft orange carpet.

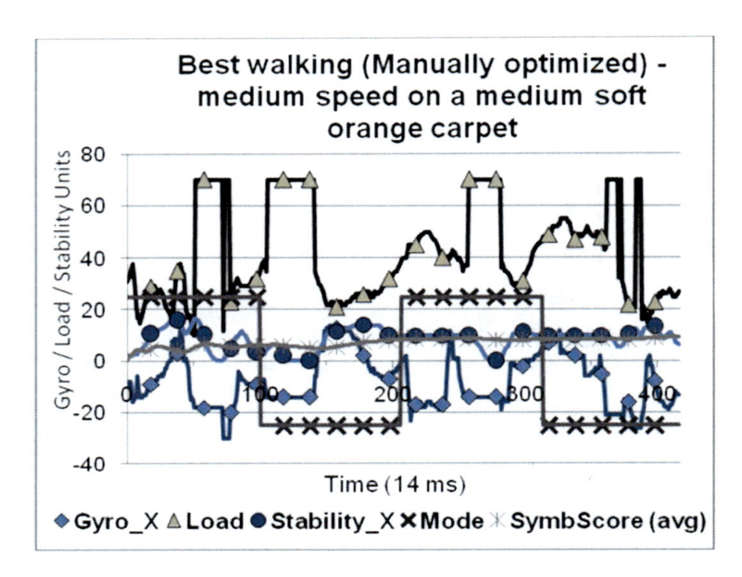

Fig. 6.46 Best walking section with "Manually optimized" walking parameters by medium robot walking speed on a medium soft orange carpet.

From the results in figures Figure 6.45, Figure 6.46, the "Load" peak values by walking parameters found with the *SelSta* approach are smaller or similar to the "Load" peak values by walking with "Manually optimized" parameters (The capped-off gyro values signalize for a overloading by servos at those particular moments of time). Also the "Gyro_X" average values are rather smaller than the "Gyro_X" average values by walking with "Manually optimized" parameters.

Fast speed

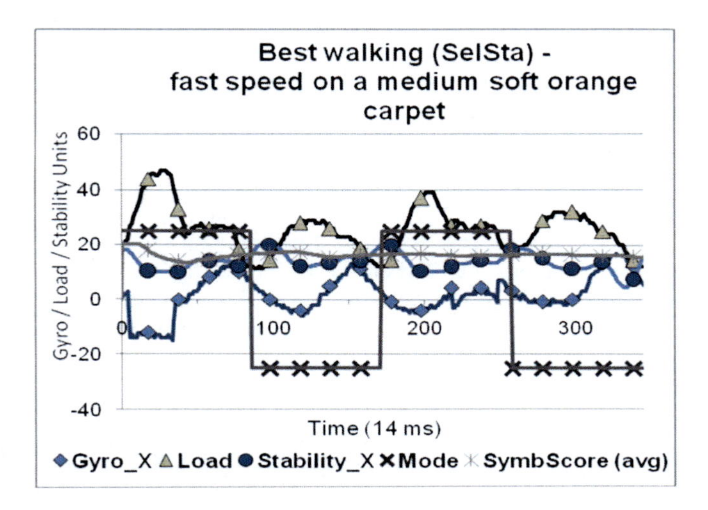

Fig. 6.47 Best walking section with SelSta approach by fast robot walking speed on a medium soft orange carpet.

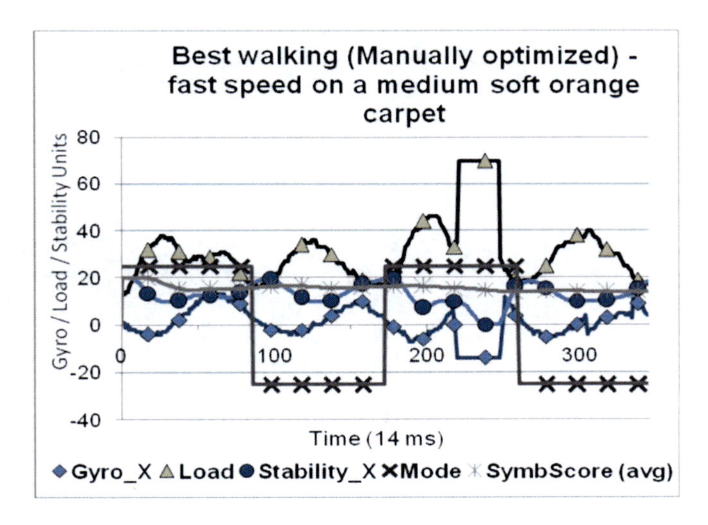

Fig. 6.48 Best walking section with "Manually optimized" walking parameters by fast robot walking speed on a medium soft orange carpet.

All this indicates that the walking parameters found with the *SelSta* approach by medium speed on a medium soft carpet can introduce more stable and perhaps more energy efficient walking than by walking with "Manually optimized" walking parameters.

From the results in figures Figure 6.47, Figure 6.48, the "Load" peak values by walking parameters found with *SelSta* approach are smaller to the "Load" peak values by walking with "Manually optimized" parameters (The capped-off gyro values signalize for a overloading by servos at those particular moments of time). Also the "Gyro_X" average values are a bit smaller than the "Gyro_X" average values by walking with "Manually optimized" parameters. All this indicates that the walking parameters found with the *SelSta* approach by medium speed on a medium soft carpet can introduce more stable and perhaps more energy efficient walking than by walking with "Manually optimized" walking parameters.

6.4.3 Experiments on a Hard Green Carpet

Experiments were done on a hard green carpet (Figure 6.49) to test out on how the SelSta approach will be able to find the optimal self-stabilizing walking on such carpet which due to its hardness and low thickness introduces dynamics in the robot's walking. Figure 6.50 shows the test setup with the S2-HuRo robot during the experiment on a hard green carpet.

Fig. 6.49 Hard green carpet.

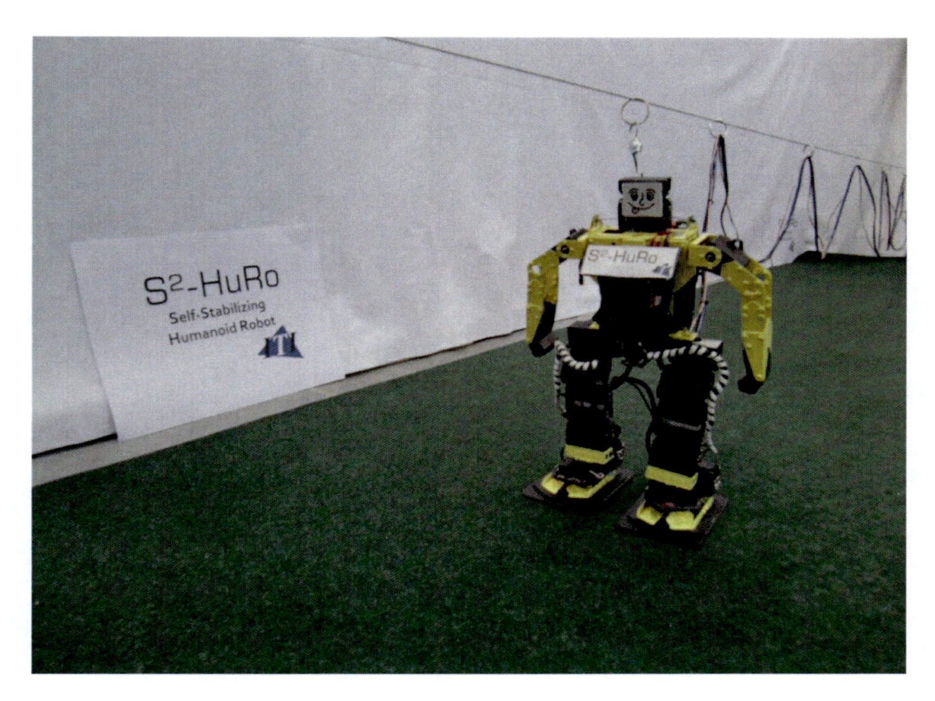

Fig. 6.50 Robot S2-HuRo on a hard green carpet.

6.4.3.1 Results from SelSta Optimal Walking Parameters Search on a Hard Green Carpet

The results from the experiments done with SelSta on the hard green carpet are divided into 3 different categories depending on the speed of the robot during the experiments: "Slow speed", "Medium speed" and "Fast speed".

Optimal walking parameters (for impulsive stabilizing movement) for slow walking speed of S2-HuRo on a medium soft carpet were found within 56 walking sections completed in 13 min. The best walking stabilization parameters found were 3.5 degrees for longitudinal and 1 degree for lateral impulsive foot movement.

Slow speed

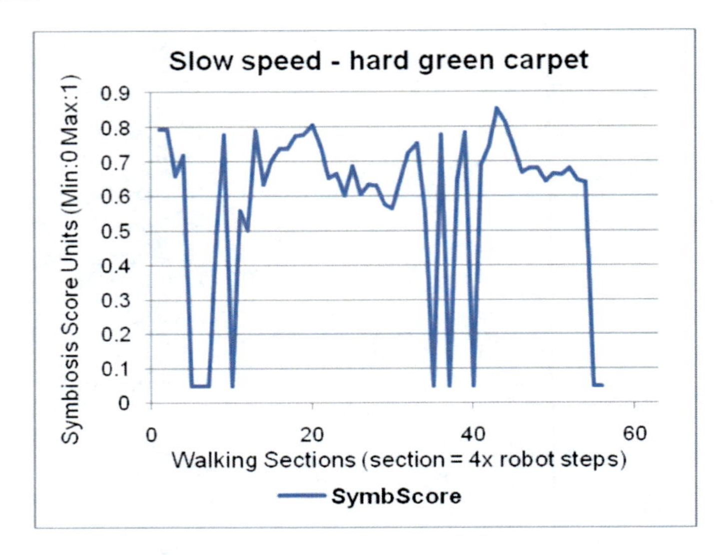

Fig. 6.51 SymbScore for slow walking speed of S2-HuRo on a hard green carpet.

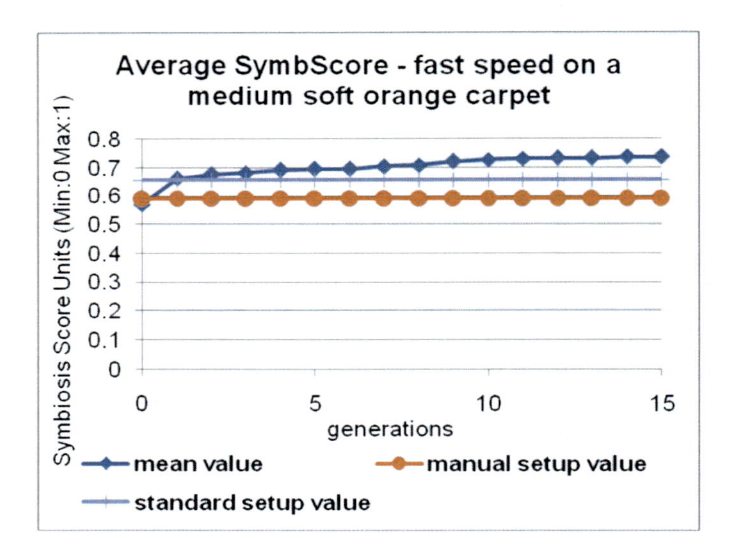

Fig. 6.52 Average SymbScore during genetic generations for slow walking speed of S2-HuRo on a hard green carpet.

Medium speed

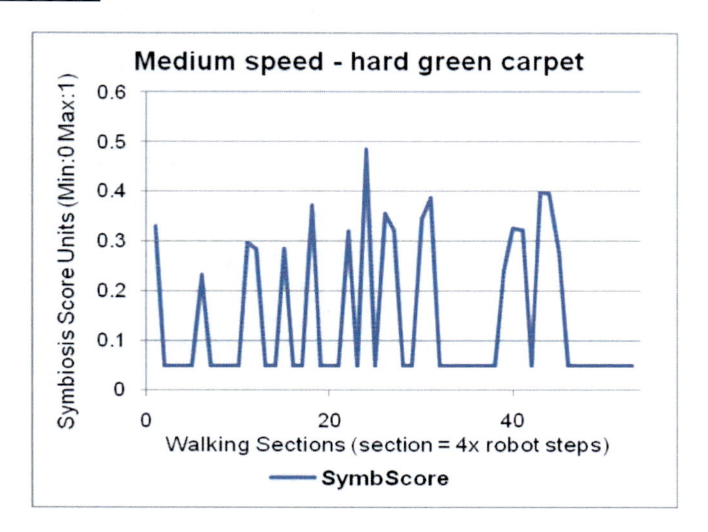

Fig. 6.53 SymbScore for medium walking speed of S2-HuRo on a hard green carpet.

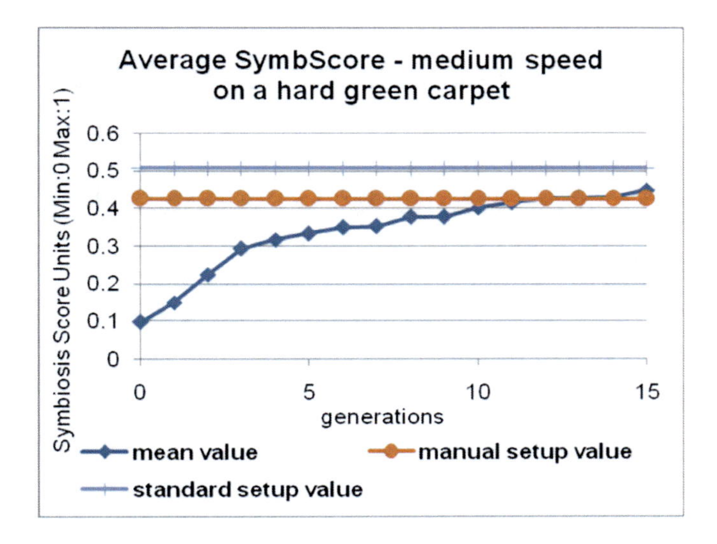

Fig. 6.54 Average SymbScore during genetic generations for medium walking speed of S2-HuRo on a medium soft orange carpet.

Optimal walking parameters (for impulsive stabilizing movement) for medium walking speed of the S2-HuRo on a medium soft carpet were found within 53 walking sections completed in 10 minutes. The best walking stabilization parameters found were 3 degrees for longitudinal and 0.5 degrees for lateral impulsive foot movement.

Fast speed

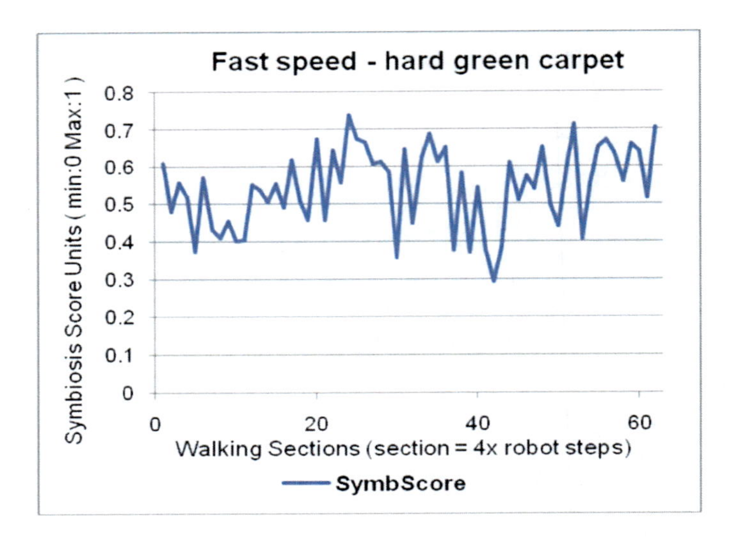

Fig. 6.55 SymbScore for fast walking speed of S2-HuRo on a hard green carpet.

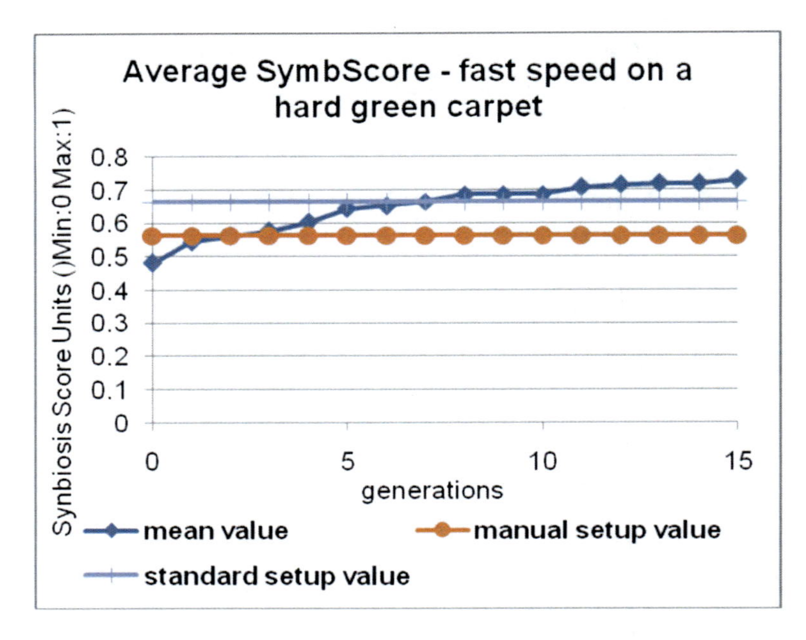

Fig. 6.56 Average SymbScore during genetic generations for fast walking speed of S2-HuRo on a hard green carpet.

Optimal walking parameters (for impulsive stabilizing movement) for fast walking speed of the S2-HuRo on a medium soft carpet were found within 62 walking sections completed in 11 minutes. The best walking stabilization parameters found were -3 degrees for longitudinal and 0.5 degrees for lateral impulsive foot movement.

6.4.3.2 Discussion about the Walking Sections Results and SymbScore Values throughout Genetic Generations by Self-stabilization on a Hard Green Carpet

From the walking sections by "Slow speed", "Medium speed" and "Fast speed" (Figure 6.51, Figure 6.53, Figure 6.55) it can be seen that by using the SelSta approach the robot fallings (the lowest *SymbScore* values on the graphs) are reduced during the experiment.

Averaged *SymbScore* (Figure 6.52, Figure 6.54, Figure 6.56) is measured over generations and direct comparison is made with the best *SymbScore* measured values found by 5 tests done with "Standard walking" parameters and by "Manually optimized walking" parameters.

For "Slow speed" and "Medium speed," after about 4 generations the *SymbScore* value is higher than the best values found by "Standard walking" and by "Manually optimized" walking parameters. Only for a medium speed, the walking with "Standard walking" parameters is better than walking with *SelSta* approach found walking parameters.

6.4.3.3 Comparison of the Best Found Values by "SelSta Walking" with the "Standard Walking", "Manually Optimized Walking" for Slow, Medium and fast Walking Speeds on a Hard Green Carpet

Once the *SelSta* approach foundt the best parameters for optimal walking with respect to the robot's stability and energy consumption while walking, 5 additional walking tests were made with those parameters and the measurements from the best walking test (with highest average *SymbScore*) were compared with the best walking test results from the 5 tests conducted with "Standard walking" (non calibrated walking parameters) and the 5 tests conducted with "Manually optimized walking" (manually calibrated walking parameters). Only the best of "Standard walking"/"Manually optimized walking" test values were selected to be compared with results from the *SelSta* approach.

As can be seen from the results in Figure 6.57 and Figure 6.58, the "Load" peak values for walking parameters found with the *SelSta* approach are relatively similar to the "Load" peak values for walking with "Manually optimized" parameters. The "Gyro_X" average values are also smaller or equal to the "Gyro_X" average values by walking with "Manually optimized" walking parameters. All this indicates that the walking parameters found with the *SelSta* approach for slow speed on a hard carpet can introduce potentially stable and energy efficient walking of equal or a better quality than walking with "Manually optimized" parameters.

Slow speed

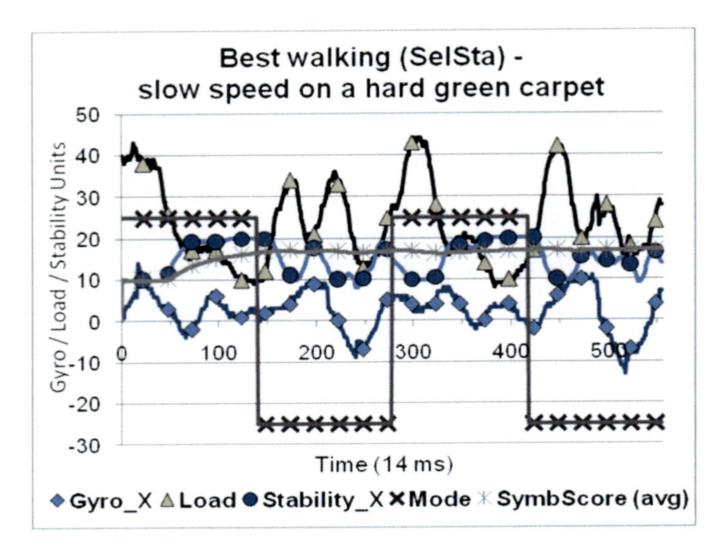

Fig. 6.57 Best walking section with SelSta approach for slow robot walking speed on a hard green carpet.

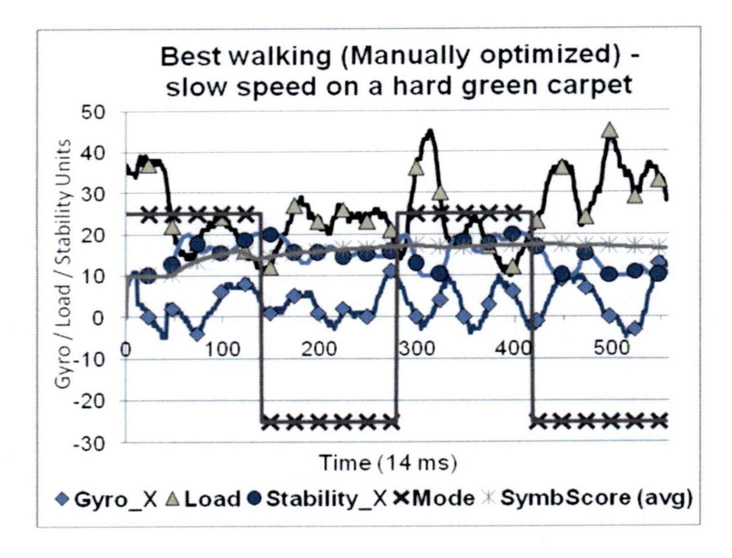

Fig. 6.58 Best walking section with "Manually optimized" walking parameters for slow robot walking speed on a hard green carpet.

From the results in Figure 6.59 and Figure 6.60, the "Load" peak values for walking parameters found with the *SelSta* approach are smaller during the robot's walking than the "Load" peak values for walking with "Manually optimized" parameters (The capped-off gyro values signify an overloading of the servos at those particular moments in time).

Medium speed

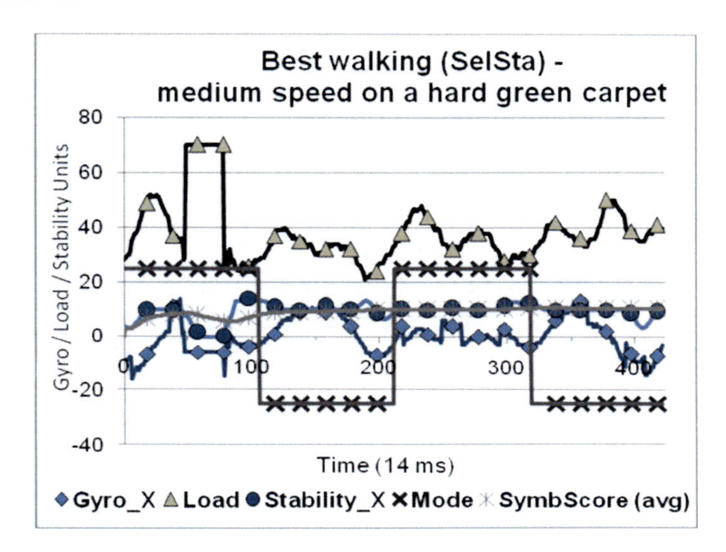

Fig. 6.59 Best walking section with SelSta approach for medium robot walking speed on a hard green carpet.

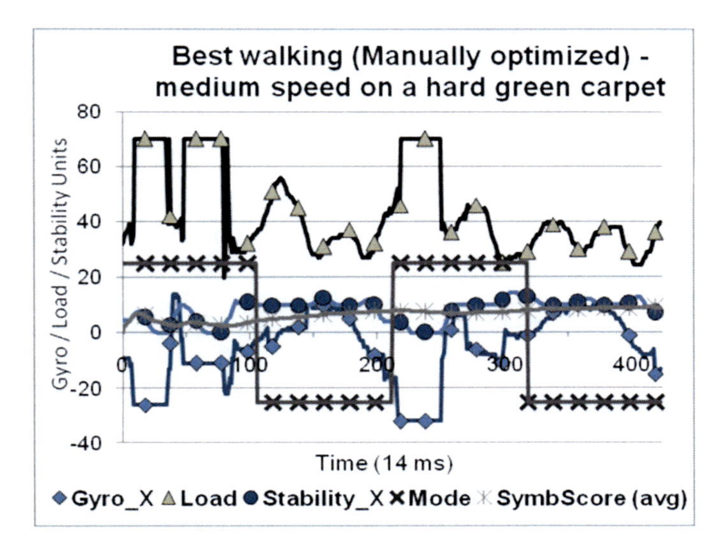

Fig. 6.60 Best walking section with "Manually optimized" walking parameters for medium robot walking speed on a hard green carpet.

Also the "Gyro_X" average values are rather smaller than the "Gyro_X" average values from walking with "Manually optimized" parameters. All this indicates that the walking parameters found with the *SelSta* approach for medium speed on a hard carpet can introduce more stable and energy efficient walking than walking with "Manually optimized" parameters.

Fast speed

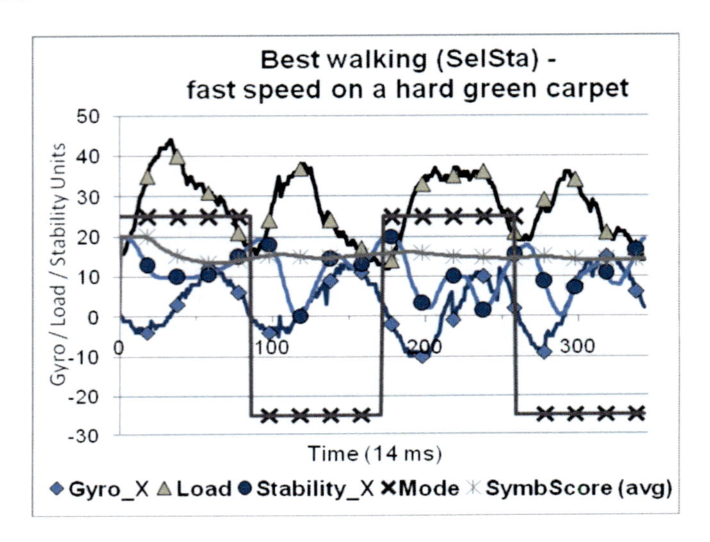

Fig. 6.61 Best walking section with SelSta approach for fast robot walking speed on a hard green carpet.

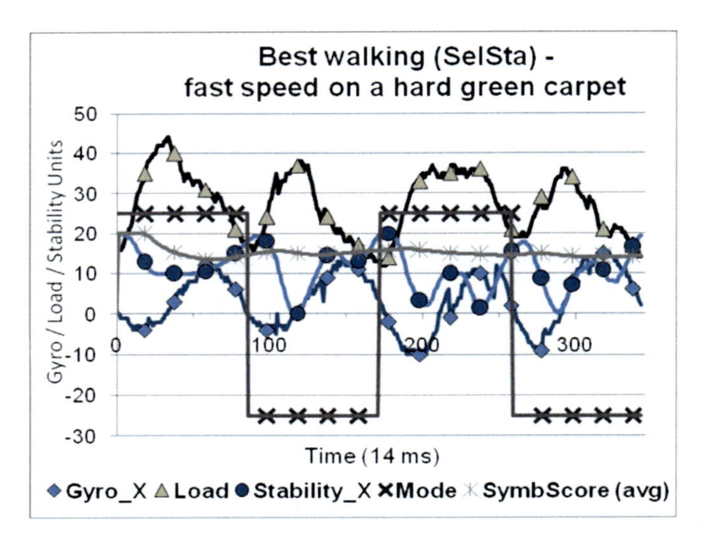

Fig. 6.62 Best walking section by "Standard walking" for fast robot walking speed over a hard green carpet.

From the results in Figure 6.61 and Figure 6.62, the "Load" peak values for walking parameters found with the *SelSta* approach are more or less equal to the "Load" peak values for walking with "Standard walking" parameters (The capped-off gyro values signify an overloading of the servos at those particular moments in time). Also, the "Gyro_X" average values are more or less equal to the "Gyro_X" average values for walking with "Standard walking" parameters. All this indicates that the walking parameters found with the *SelSta* approach for fast speed on the hard carpet can introduce more stable and energy efficient walking than by "Standard walking" parameters.

6.4.4 Experiments on a Hard Linoleum Surface

Experiments were also done on a hard linoleum surface (Figure 6.63) to test on how well the SelSta approach would be able to find the optimal self-stabilizing walking on such hard surface which due to its hardness introduces dynamics in the robot's walking. Figure 6.64 shows the test setup with the S2-HuRo robot during the experiment on a hard linoleum surface.

Fig. 6.63 Hard linoleum surface

6.4.4.1 Results from SelSta Optimal Walking Parameters Search on a Hard Linoleum Surface

The results from the experiments done with SelSta on a hard hard linoleum surface are divided into 3 different categories depending on the speed of the robot during the experiments: "Slow speed", "Medium speed" and "Fast speed".

Fig. 6.64 Robot S2-HuRo on a hard linoleum surface.

Slow speed

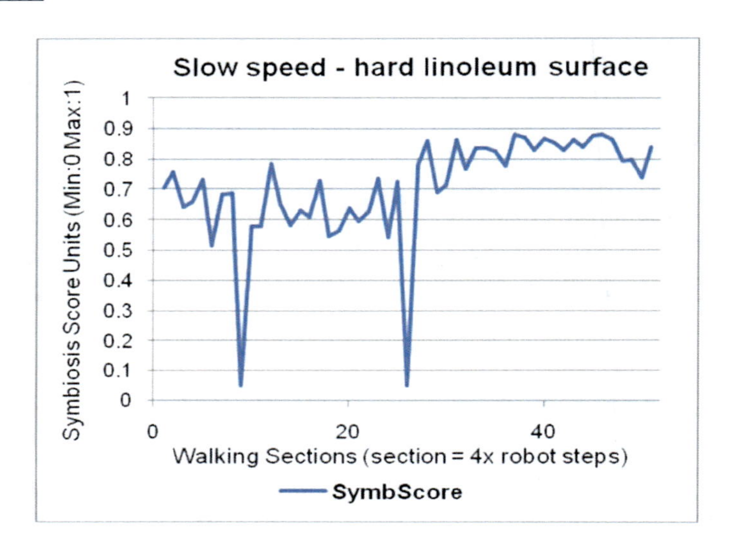

Fig. 6.65 SymbScore for slow walking speed of S2-HuRo on a hard linoleum surface.

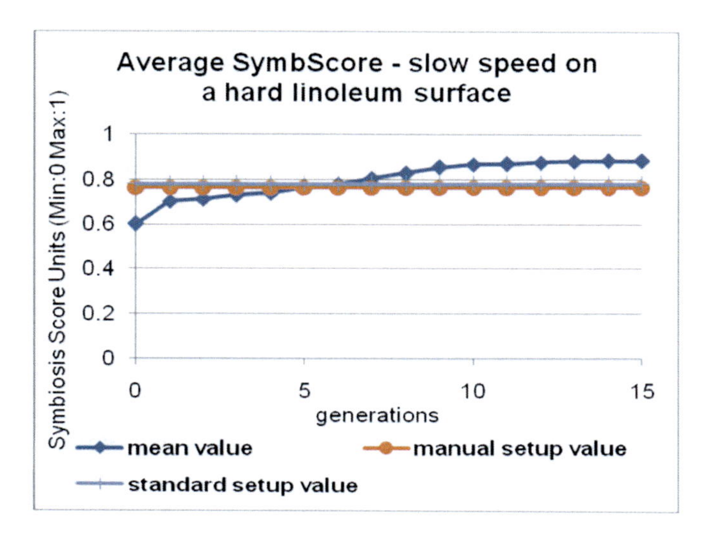

Fig. 6.66 Average SymbScore during genetic generations for slow walking speed of the S2-HuRo on a hard linoleum surface.

Optimal walking parameters (for impulsive stabilizing movement) for slow walking speed of the S2-HuRo on hard linoleum were found within 51 walking sections completed in 11 minutes. The best walking stabilization parameters found were -0.5 degrees for longitudinal and -0.5 degree for lateral impulsive foot movement.

Medium speed

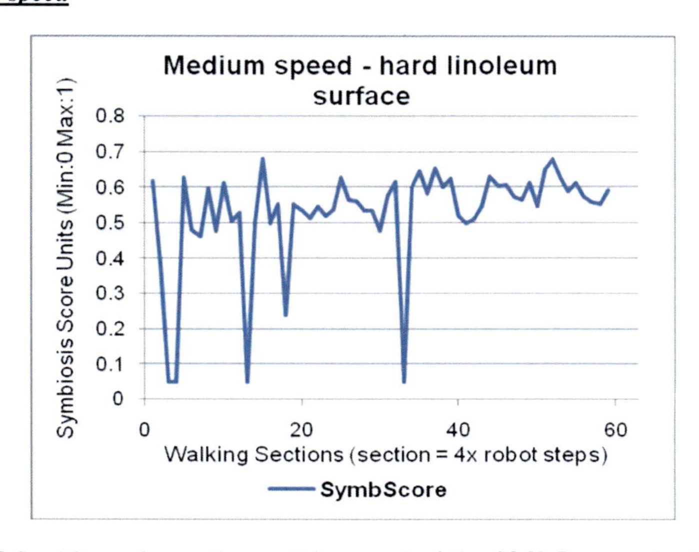

Fig. 6.67 SymbScore for medium walking speed of the S2-HuRo on a hard linoleum surface.

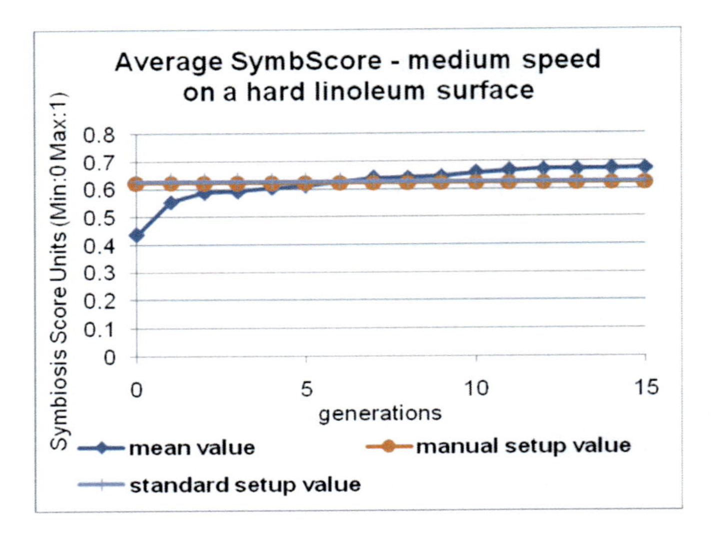

Fig. 6.68 Average SymbScore during genetic generations for medium walking speed of the S2-HuRo on a hard linoleum surface.

Optimal walking parameters (for impulsive stabilizing movement) for medium walking speed of the S2-HuRo on hard linoleum were found within 59 walking sections completed in 12 minutes. The best walking stabilization parameters found were 2.5 degrees for longitudinal and -0.5 degrees for lateral impulsive foot movement.

Fast speed

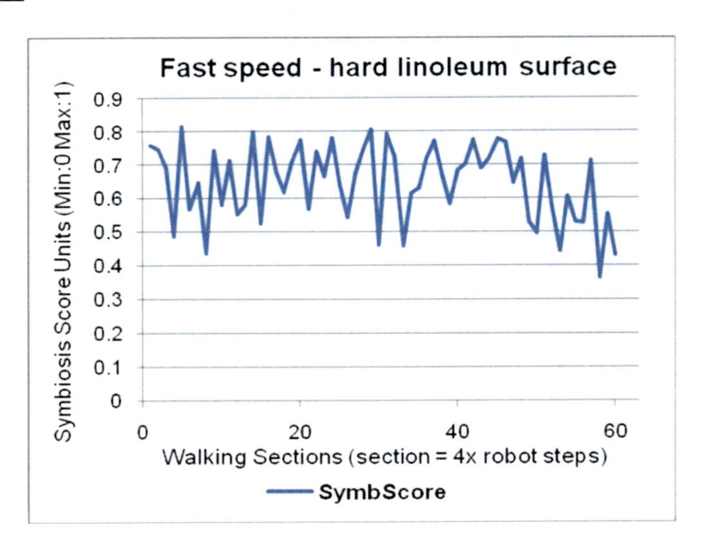

Fig. 6.69 SymbScore for fast walking speed of the S2-HuRo on a hard linoleum surface.

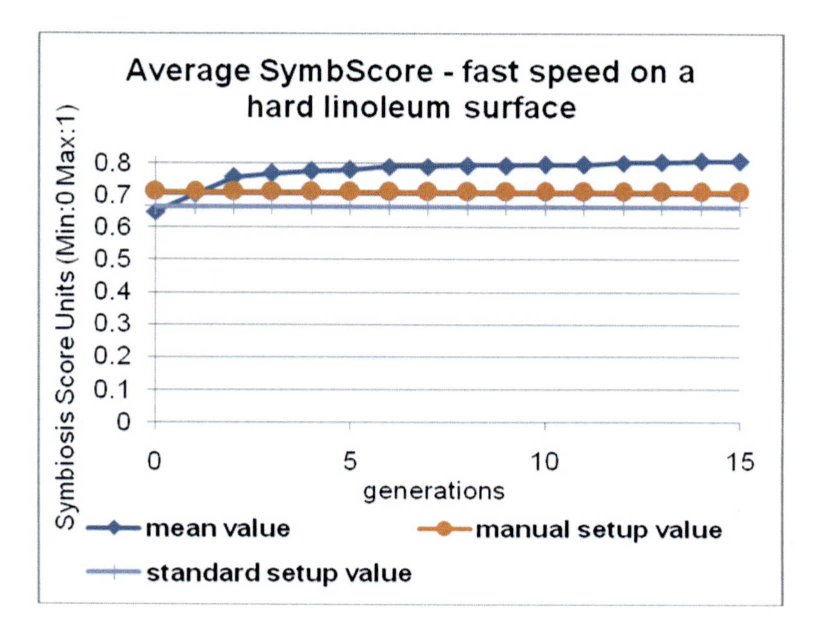

Fig. 6.70 Average SymbScore during genetic generations for fast walking speed of the S2-HuRo on a hard linoleum surface.

Optimal walking parameters (for impulsive stabilizing movement) for fast walking speed of the S2-HuRo on hard linoleum were found within 60 walking sections completed in 11 minutes. The best walking stabilization parameters found were 2.5 degrees for longitudinal and -1 degree for lateral impulsive foot movement.

6.4.4.2 Discussion about the Walking Sections Results and SymbScore Values throughout Genetic Generations for Self-stabilization on a Hard Linoleum Surface

From the walking sections for "Slow speed", "Medium speed" and "Fast speed (Figure 6.65, Figure 6.67, Figure 6.69) it can be seen that by using the SelSta approach the robot fallings (the lowest *SymbScore* values on the graphs) are reduced during the experiment.

Averaged *SymbScore* (Figure 6.66, Figure 6.68, Figure 6.70) measured over generations and a direct comparison is made with the best *SymbScore* measured values found by 5 tests done with "Standard walking" parameters and by "Manually optimized walking" parameters.

Only for medium speed after about 10 generations, the *SymbScore* value is higher than the best values found by "Standard walking" and by walking with "Manually optimized" walking parameters.

6.4.4.3 Comparison of the Best Found Values by "SelSta Walking" with the "Standard Walking", "Manually Optimized Walking" for Slow, Medium and Fast Walking Speeds on a Hard Linoleum Surface

Once the *SelSta* approach found the best parameters for optimal walking with respect to robot's stability and energy consumption during walking, 5 additional walking tests were made with those parameters and the measurements from the best walking test (with highest average *SymbScore*) were compared with the best walking test results from the 5 tests conducted with "Standard walking" (non calibrated walking parameters) and the 5 tests conducted with "Manually optimized walking" (manually calibrated walking parameters). Only the best values from "Standard walking"/"Manually optimized walking" were selected to be compared with the *SelSta* approach.

As can be compared from the results in Figure 6.71 and Figure 6.72, the "Load" peak values for walking parameters found with *SelSta* approach are a bit greater than during the robot's walking than the "Load" peak values from walking with "Manually optimized" parameters. The "Gyro_X" average values are smaller or equal to the "Gyro_X" average values from walking with "Manually optimized" parameters.

Slow speed

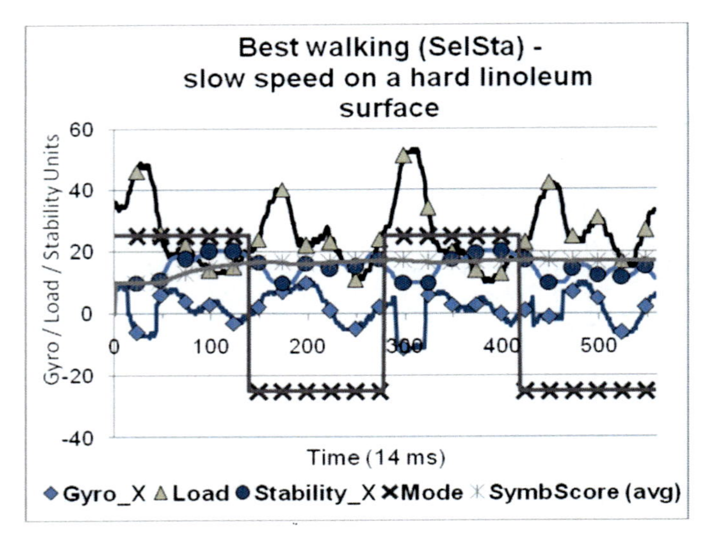

Fig. 6.71 Best walking section with SelSta approach for slow robot walking speed on a hard linoleum surface.

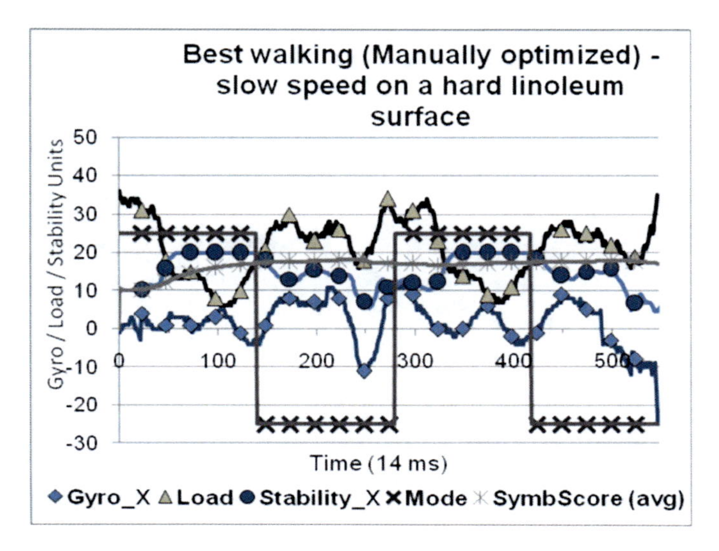

Fig. 6.72 Best walking section with "Manually optimized" walking parameters for slow robot walking speed on a hard linoleum surface.

All this indicates that the walking parameters found with the *SelSta* approach for slow speed on the hard linoleum surface can introduce potentially equally stable and energy efficient walking as walking with "Manually optimized" parameters.

Medium speed

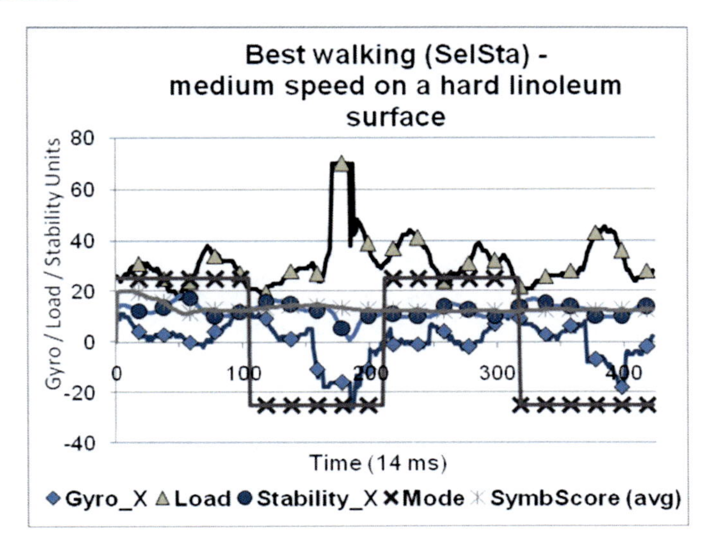

Fig. 6.73 Best walking section with SelSta approach for medium robot walking speed on a hard linoleum surface.

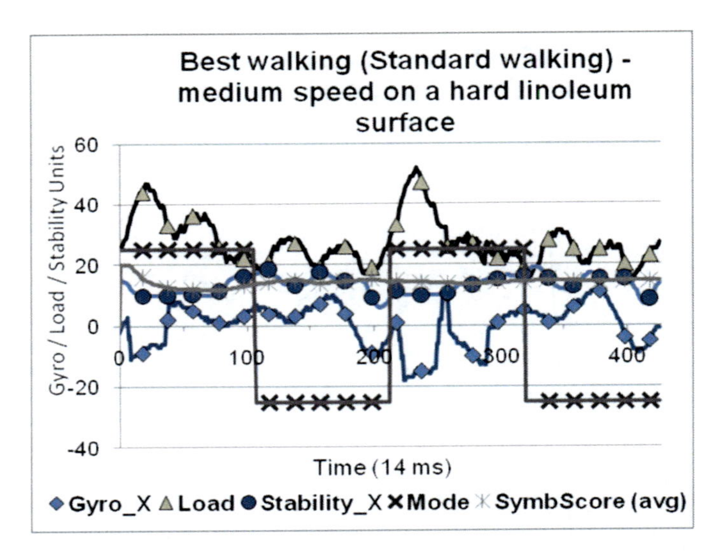

Fig. 6.74 Best walking section by "Standard walking" for medium robot walking speed on a hard linoleum surface.

From the results in Figure 6.73 and Figure 6.74, the "Load" peak values for walking parameters found with the *SelSta* approach are smaller or equal to the "Load" peak values for walking with "Normal walking" parameters (The capped-off gyro values signify an overloading of the servos at those particular moments in time). The "Gyro_X" average values are smaller than the "Gyro_X" average values for walking with "Standard walking" parameters. All this indicates that the walking parameters found with the *SelSta* approach for medium speed on the hard linoleum can introduce more stable walking than by walking with "Standard walking" parameters.

Fast speed

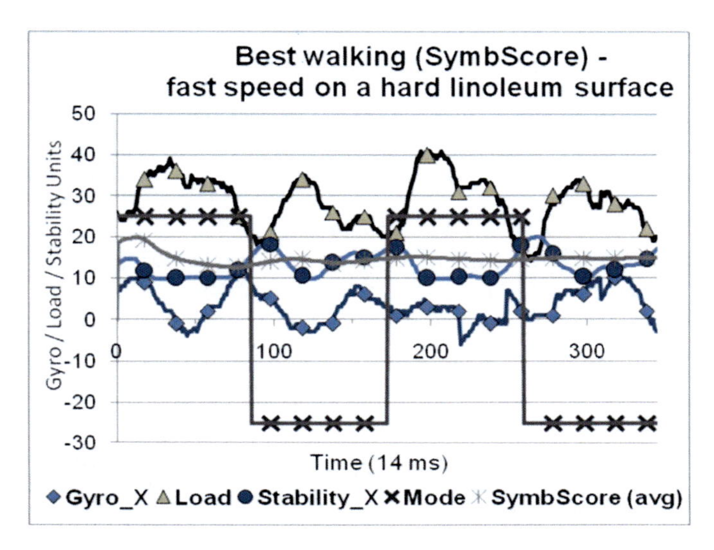

Fig. 6.75 Best walking section with SelSta approach for fast robot walking speed on a hard linoleum surface.

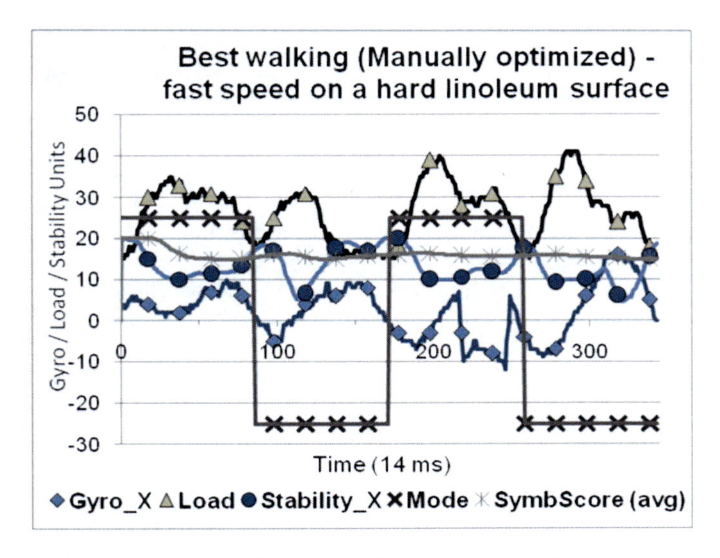

Fig. 6.76 Best walking section with "Manually optimized" walking parameters for fast robot walking speed on a hard linoleum surface.

From the results in Figure 6.75 and Figure 6.76, the "Load" peak values for walking parameters found with the *SelSta* approach are more or less equal to the "Load" peak values for walking with "Manually optimized" parameters. The

"Gyro_X" average values are smaller than the "Gyro_X" average values for walking with "Manually optimized" parameters. This indicates that the walking parameters found with the *SelSta* approach for fast speed on the hard linoleum surface can introduce more stable walking compared to walking with "Manually optimized" parameters.

6.5 Summary for Experiments Done with the SelSta Approach

The results of the experiments are summarized in Figure 6.77. They include data from performance tests done on 5 walking test sections (each section is 6 robot walking steps as described earlier) of the best balancing parameters found using *SelSta*. They also give a direct comparison to tests done with manually set values for the balancing parameters and the standard set values (foot longitudinal and lateral degree set to 0). The comparison also includes the number of robot fallings on final test runs.

Surface	Green Carpet (Hard)																	
Method	Autonomous Self-Stabilizing						Manual Values						Standard Values					
Walking Speed	Slow		Medium		Fast		Slow		Medium		Fast		Slow		Medium		Fast	
Foot axis (X-Longitudinal; Y-Lateral)	X	Y	X	Y	X	Y	X	Y	X	Y	X	Y	X	Y	X	Y	X	Y
Stabilization parameters (axis degrees)	3.5	1	3	0.5	0.5	-3	1	4	1	4	1	4	0	0	0	0	0	0
Self-Stabilization approach duration	13 min		10 min		11 min		10-14 hrs						/					
SymbScore (average of 5 test sections)	0.8		0.51		0.66		0.72		0.42		0.56		0.69		0.11		0.63	
SymbScore (best of 5 test sections)	0.85		0.53		0.72		0.82		0.46		0.65		0.82		0.34		0.71	
Robot fallings (in 5 test sections)	0		0		0		0		0		0		0		3		0	

Surface	Green Carpet (Soft)																	
Method	Autonomous Self-Stabilizing						Manual Values						Standard Values					
Walking Speed	Slow		Medium		Fast		Slow		Medium		Fast		Slow		Medium		Fast	
Foot axis (X-Longitudinal; Y-Lateral)	X	Y	X	Y	X	Y	X	Y	X	Y	X	Y	X	Y	X	Y	X	Y
Stabilization parameters (axis degrees)	1.5	0.5	2	4	0.5	-4.5	1	4	1	4	1	4	0	0	0	0	0	0
Self-Stabilization approach duration	16 min		14 min		12 min		10-14 hrs						/					
SymbScore (average of 5 test sections)	0.8		0.28		0.65		0.72		0.3		0.54		0.73		0.22		0.6	
SymbScore (best of 5 test sections)	0.85		0.56		0.74		0.79		0.42		0.58		0.83		0.38		0.62	
Robot fallings (in 5 test sections)	0		1		0		0		1		0		0		0		0	

Surface	Linoleum Surface (Hard)																	
Method	Autonomous Self-Stabilizing						Manual Values						Standard Values					
Walking Speed	Slow		Medium		Fast		Slow		Medium		Fast		Slow		Medium		Fast	
Foot axis (X-Longitudinal; Y-Lateral)	X	Y	X	Y	X	Y	X	Y	X	Y	X	Y	X	Y	X	Y	X	Y
Stabilization parameters (axis degrees)	-0.5	-0.5	2.5	-0.5	2.5	-1	1	4	1	4	1	4	0	0	0	0	0	0
Self-Stabilization approach duration	11 min		12 min		11 min		10-14 hrs						/					
SymbScore (average of 5 test sections)	0.82		0.62		0.72		0.76		0.62		0.71		0.78		0.61		0.66	
SymbScore (best of 5 test sections)	0.86		0.64		0.75		0.85		0.71		0.75		0.81		0.73		0.73	
Robot fallings (in 5 test sections)	0		0		0		0		0		0		0		0		0	

Surface	Orange Carpet (Medium Soft)																	
Method	Autonomous Self-Stabilizing						Manual Values						Standard Values					
Walking Speed	Slow		Medium		Fast		Slow		Medium		Fast		Slow		Medium		Fast	
Foot axis (X-Longitudinal; Y-Lateral)	X	Y	X	Y	X	Y	X	Y	X	Y	X	Y	X	Y	X	Y	X	Y
Stabilization parameters (axis degrees)	1	2.5	1.5	4	1.5	0	1	4	1	4	1	4	0	0	0	0	0	0
Self-Stabilization approach duration	16 min		14 min		11 min		10-14 hrs						/					
SymbScore (average of 5 test sections)	0.68		0.36		0.72		0.63		0.24		0.59		0.58		0.09		0.65	
SymbScore (best of 5 test sections)	0.77		0.57		0.78		0.72		0.46		0.71		0.67		0.27		0.7	
Robot fallings (in 5 test sections)	0		1		0		0		2		0		0		4		0	

Fig. 6.77 Results from humanoid robot self-stabilizing experiments done on different kinds of surfaces, with different testing parameters and three different walking speeds.

As can be seen from the results in the table, the *SelSta* approach has clearly reached its projected goals. In comparison with manual setup values, the self-stabilizing *SelSta* approach generates a more stable, energy efficient walking of the humanoid robot on different kinds of flat surfaces in relatively short time.

In the table, the best results from the tests are shown in lime green. Only two test runs show the "Manually optimized" and "Standard parameters" producing slightly better results than the *SelSta* found walking parameters.
It can be noticed that the stability on the soft carpet can be considerably improved by using the *SelSta* approach. For fast walking speed the stability of the robot can be significantly improved by using the *SelSta* approach, whereas for medium walking speed this improvement is not that big.

In the analysis of the tests and results from the test sections it can be seen that with parameters found by the *SelSta* approach there is a smaller overload and the walking of the robot is also more stable. Reducing the overload has big influence on energy consumption as well the life of the servo elements and electronic components.

The fallings from the *SelSta* approach walking parameters are smaller in comparison to walking using the "Manually optimized" and "Standard parameters".

The walking parameters found with the *SelSta* approach are usually found within 11 to 16 minutes. For fast walking speed of the robot, the *SelSta* approach needed 11 to 12 minutes for finding the optimal walking parameters, which is far less than the 10 to14 hour it takes for manually optimizing the walking parameters for each walking robot speed for each particular flat surface.

The *SelSta* approach demonstrates that *self-stabilizing* optimal humanoid robot walking on a flat surface can be achieved using a biologically inspired approach.

Chapter 7
Biologically Inspired Approaches for Anomaly Detection within a Robotic System

Fault tolerance is an important characteristic for autonomous robotic systems that operate and fulfill mission tasks in an autonomous way. Being fault-tolerant means the robot is:

- more reliable in its operation;
- can withstand some faulty situations that may occur during its mission;
- can mitigate such failures without a human operator;
- can continue implementing its mission assignments.

Making robotic systems fault-tolerant is not often easily achieved however. This is often due to the fact that robotic systems consist of various hardware and software parts in which faults may occur. Therefore the fault tolerant robot should be able to cope with its mechanical, electronic, and software malfunctions.

Therefore emphasis is put on developing a fault detection approach with characteristics such as on-line learning and situation adaptation, since those features will enable the robot to operate in a more autonomous way. In this chapter an interdisciplinary research is presented on developing a robust algorithmic approach for anomaly detection for the domain of self-capable robots. Self-capable robots are robots that poses some of the self-x properties, such as: self-reconfiguration, self-adaptation, self-optimization, self-stabilization, etc.

Before describing the approach further in detail, a comparison on common fault detection techniques used in the robotics domain is presented and their properties explained.

7.1 Overview on Approaches for Fault / Anomaly Detection by Robotic Systems

Research on developing fault tolerant approaches for general fault-tolerant real time systems can often be associated with integration of redundant systems running identical electronics and software. In case of failure of an electronic computing component, the failed component can be detected and isolated by

B. Jakimovski: Biologically Inspired Approaches, COSMOS 14, pp. 127–150.
springerlink.com

means of voting of redundant systems. Although robots are often considered real time systems, they don't usually have additional redundant components, sensors, and actuators because this would increase their size (the size of the robot is often critical for particular tasks that it can execute) as well as their price for development and even further increasing the complexity of the system and the energy consumption. Namely robotic systems are usually many degrees of freedom from embedding many sensors and actuators, therefore complex enough even without additional redundant systems on board. The complexity has influence on the modeling of the robotic system, not just for its control but also for the model based fault detection and isolation within such a system. However, building a full mathematical model for a robotic system and fault detection can often be a tedious task [MAV09] [LiC05] [MDD04] [DeM05] since not all of the details and circumstances can be pre-modeled by a human operator. On the other hand, modeling of fault detection for a complex robotic system also diverges from the initial idea of having self-capable robotic systems, where the human designer effort for maintenance should be minimal.

Other approaches such as: Time-Delay Neural Networks (TDNN) [COB08], Recurrent Neural Networks (RNN) [Prz06], and particle filters [VeS06] [ZMZ06] have been considered for fault detection which are perhaps more robust than the classical modeling approaches and require less human effort to be implemented and maintained. The mentioned approaches are mostly related to the procedure of synthesizing fault detection components based on the data collected during training runs. After the fault detectors have been synthesized they are evaluated on fault scenarios in real time, or on the recorded data, and the fault positives or fault negatives are estimated. However, generating fault detectors for every possible fault scenario that can occur is sometimes an exhaustive process.

Therefore research has been made here on developing a self-adapting anomaly detection approach that adapts to the environment and situations in which the robot is operating, similar to biological processes. Being self-adapting, the anomaly detection would also need less human effort for developing and maintaining of such an anomaly detection mechanism for a particular robotic system – which is one of the characteristics an autonomous robot should possess.

7.2 Overview of Artificial Immune System (AIS) Concept

For developing a self-adapting anomaly detection approach for the robotics domain, inspiration was found in processes seen in immune systems (Figure 7.1). Namely the immune systems of vertebrates are related to detection and removal of pathogens and their main property is to distinguish between the bacteria, viruses, and the organism's own cells. In the figure Figure 7.1 it is represented how the B-cells interact with the antigens and the production of antibodies and memory cells that "remember" the antigen cell patterns.

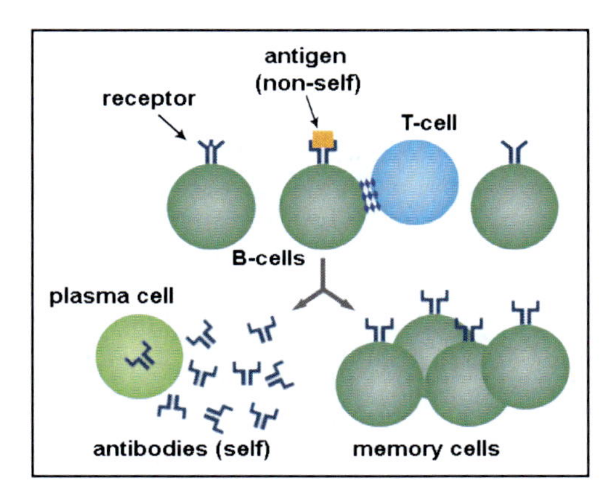

Fig. 7.1 Antibodies(self), antigens(non-self) and other cell types of the immune system.

Artificial Immune System (AIS) is a computational paradigm that is related to properties of the immune systems and these concepts are applied in problem solving [DeJ02] in various engineering domains. One of the main AIS properties are detection and memorization, similar to how the immune system of living organisms detects and memorize the specific patterns of pathogens.

The AIS has been successfully applied to various domains: pattern recognition [CaD03], data mining [NCR05], network security [PaV05], robotics domain [MiV02], [SiN05], [SaS02], [NFR06], [CJT03], and others.

The motivation behind using the AIS concept for developing the self-adapting anomaly detection approach was due to the anomaly detection process of the robot can be compared to the process of detection of pathogens seen by the immune system. Additionally, other AIS properties have been researched so the self-adapting property of the newly introduced anomaly detection system can be realized.

There are several techniques related to AIS, which are based on immunological theories on how the immune system functions:

- Negative selection;
- Positive selection;
- Immune networks;
- Dendritic Cell algorithms.
- Clonal Selection;

Negative selection mechanism is based on the ability of the immune system to learn to distinguish between non-self cells and self cells. The aim of negative selection is to provide tolerance of the self cells. As noted in [FPA94], negative selection consists of generating a set of detectors and evaluation of those detectors. Only the detectors that do not react to self are considered for further detection.

Positive selection is based on enabling detectors that can detect non-self cells.

In *immune networks*, a property of the immune system is taken into consideration, where antibody cells (self) have a mechanism for stimulating and suppressing each other in an idiotypic network. The antibody cells are connected between themselves only if the affinities between them exceed a set threshold level.

Dendritic Cell algorithms are based on an abstract model of dendritic cells and behavior exhibited by the population of the cells. (Dendritic cells are immune cells that form part of the mammalian immune system. Their main function is to process antigen material and present it on the surface to other cells of the immune system.)

Clonal selection method is based on a mechanism that antibody cells (self) implement upon recognizing the antigen cells (non-self). Upon discovery of an antigen, antibodies start to proliferate by cloning themselves. They also memorize the antigenic attack (immune memory) so they have higher responsiveness on the particular antigenic attack over time. (This is similar to primary and secondary immune response - Figure 7.2).

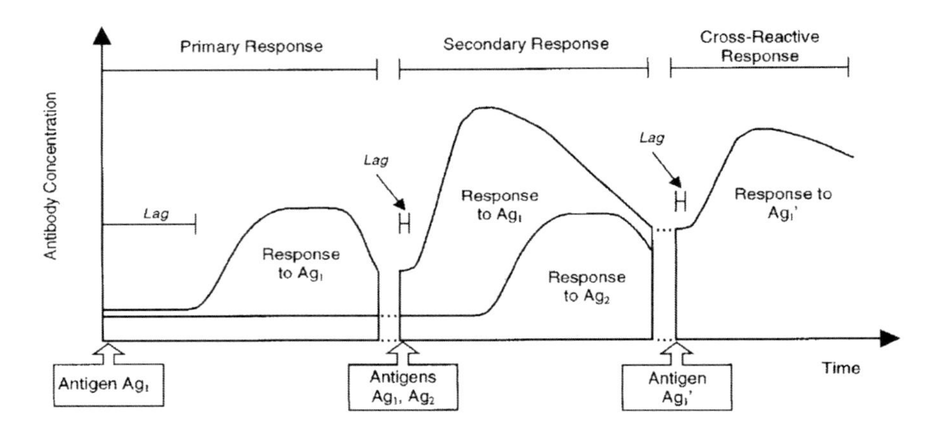

Fig. 7.2 Primary and secondary immune response [DeV02].

Clonal selection is illustrated in Figure 7.3.

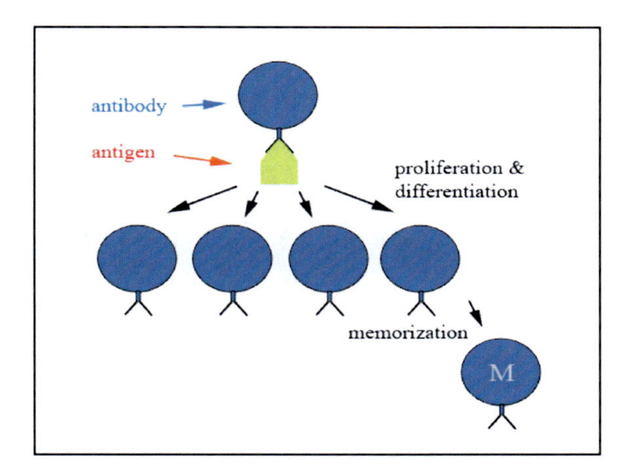

Fig. 7.3 Illustration of Clonal proliferation.

7.3 Artificial Immune System Based - Robot Anomaly Detection Engine (RADE) Approach

Various artificial immune system approaches have been considered for the robotics domain such as: negative selection [NFR06], inflammation [SaS02], etc.

An AIS concept based on clonal selection principle was used for development of a self-adaptable anomaly detection algorithm named Robot Anomaly Detection Engine (RADE) [JaM08]. The clonal selection principle was chosen since the dynamics of proliferation of antibodies (self) can be useful for describing the runtime dynamics of the anomaly detection process. The clonal selection principle in RADE was combined with another non linear computational method – fuzzy logic. Fuzzy logic is multi-valued logic related to fuzzy set theory [Zad65] that deals with imprecise reasoning. The fuzzy logic approach was chosen for information representation in RADE since an exact recognition for particular non-self is not necessary to trigger an anomaly response. This is similar to the immune's system functioning, where by a given similar (but not necessarily identical) stimulus, the immune system response can be initiated. Additionally, fuzzy logic rules maximize the imprecision in defining them and on the other hand minimize the factor of predefining the possible faulty states in details.

For RADE, the fuzzy logic rules are related to describing the self / non-self situations which are categorized into self / non-self sets. The rules in the non-self set detect when there is an anomaly present within the system, and the rules in the self set detect when the situation is not characterized as anomalous. For example, non-anomalous situations can be interpreted in relation to the sensor or actuator values. An example of this would be that walking behavior should not have a low acceleration level; or walking with an anomaly should cause the servo's current to be high. The fuzzy logic rules in self / non-self sets describe such similar situations with IF – THEN fuzzy rules. Those rules in RADE have the following structure:

IF *X1&X2&...Xn* THEN *Y* WITH *weight_factor_Z*

The "*X1&X2&...Xn*" represents the premise part which consists of monitored behaviours such as walking, standing, etc.; and some particular characteristics like current, acceleration, etc. The "*Y*" is the consequent part and can have two types of values: "anomaly is present" or "anomaly is absent" (Figure 7.4). The weight factor "*Z*" represents the clonal proliferation within AIS, and is in a range from 0.0 to 1.0. The "*Z*" value will increase for some constant value (for example: 0.1) if the rule has "fired" by fuzzy logic evaluation. In parallel to that the weights will decrease in all the rules belonging to the opposite set, just as the concentration of self/non-self drops down being influenced by an increased concentration of non-self/self within the immune system. The firing level of each rule is therefore always adjusted, depending on the value of "*Z*".

Here is an example of such rules:

IF Current is Very_High THEN Anomaly is Present WITH 0.3; - belonging to the non-self set

or

IF Current is Very_Small THEN Anomaly is Absent WITH 0.3; - belonging to the self set

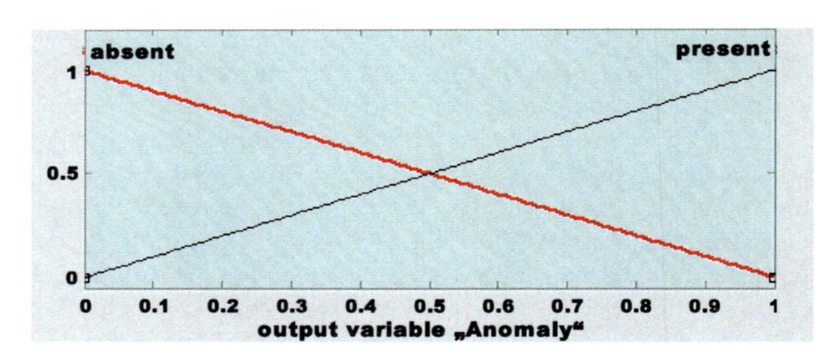

Fig. 7.4 Fuzzy membership set for the output "Anomaly" variable.

The premise is related to a situation when the current (of a servo for example) has a very high value, while the "Anomaly is Present" indicates that the situation belongs to the non-self set, i.e. indicating the possibility of an anomaly within the system. The 0.3 at the end of the rule is for example implementation of the weight factor "*Z*" in that particular moment in time. The weight factor "*Z*" also has another positive characteristic for the anomaly detection engine. Namely the idea is to reduce the factor of hand coded elements in RADE, and let the system dynamically adjust itself to the current situation, can be characterized as self-adaptation.

The importance of usage of the weight factor "Z" can be explained as follows:

In case we have fuzzy rules coded without using weights and depending on the manually pre-designed fuzzy membership sets for the premise parts of the rules, the rules can have an optimal response for a given situation and perhaps an unsatisfactory response for other unforeseen situations.

By the fuzzy rules with weights, this would introduce two additional features:

- The premise parts of the rules do not require any additional handcrafting and expert designing for their fuzzy membership sets. Therefore they can have some automatically generated "standard" triangular fuzzy membership sets, normally distributed within a valid range for the observed variable. For example, such fuzzy membership sets the monitored variable "Current" to having values in the range from min 0 to max 3 Amperes. This can be seen as in Figure 7.4.

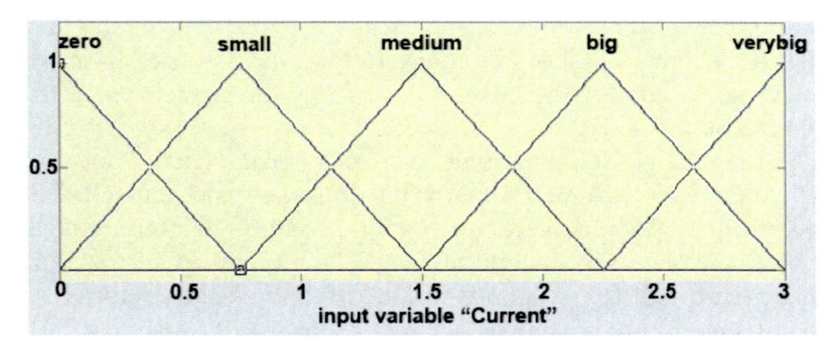

Fig. 7.5 Fuzzy membership set for monitored variable "Current".

  The other membership sets for other monitored variables (such as acceleration and temperature) are going to have the same "standard" five or more triangular fuzzy membership sets normally distributed for their input range. An interesting aspect about having such "standard" fuzzy membership sets is that they can be part of the learning process, which is to be introduced in the next step of the development of RADE. Having the "standard" fuzzy membership sets for each of the observed parameters, a rule-based learning system can be built which incorporates only new situations described by new rules. This is effective since the fuzzy membership sets for the observed parameters are not going to be changed, and it will be possible to distinguish between what has already been learned and what can be learned.

- The weight factors for such rules having the "standard" generated fuzzy membership sets allows the rule to adapt to the situation even without the rule being optimally pre-designed at the start, i.e. having its membership sets optimally pre-designed for that particular situation. Therefore the changes in the weights depend on the particular situation and therefore contribute to the *self-adapting* dynamics of overall anomaly detection system.

7.3.1 Core Functionality of RADE Approach

The functioning principle of RADE, the self and non-self rule sets, and the dynamical change of the weights by those sets can be further explained with an example situation presented in Figure 7.6. In order to concentrate on core properties of the functioning principle of RADE, in Figure 7.6 only the behaviour "Walking" and the monitored parameter "Current" have been presented, although in some other situations also other parameters such as "Acceleration" and "Temperature" in correlation with other behaviours like "Climbing" and "Descending" can be observed.

As illustrated in the figure, when the premise within a rule belonging to the non-self set is satisfied, the rule "fires" (in fuzzy logic interpretation) and its weight is also increased by some constant value, e.g. by 0.1 from 0.3 to 0.4. At the same time, the weights of rules belonging to the other set are all decreased with the same constant value of 0.1. In such a way they lower their value from 0.4 on 0.3 or from 0.6 on 0.5 and so on.

In each computation step, a weighted output is calculated from such a fuzzy system which contains two membership functions: self and non-self. The value is in the range from 0.0 (no anomaly) to 1.0 (full anomaly) and represents the output of the RADE method. The output of RADE is computed in a defuzzification process as a centroid of fuzzy outputs of the "fired" rules. Therefore the output of RADE is influenced by the weight factor of each of the firing rules. The weight factor acts in a similar way as the secondary and subsequent responses within the immune system, i.e. the more the weight is associated to some rules, the more significant the response of those rules will be for the output of RADE in the next case of firing.

The change of weights therefore acts as some sort of *short term memory* for events that occurred some moments ago. The anomaly output level of RADE depends on its short history and also on the actual system's state. S*hort term memory* is also an important characteristic of RADE's functioning, since by using this principle some intermediate "spikes" in sensors readings can have less influence on the computed health signal value. Health signal value is within a range of valid values where the lowest min value is some completely healthy state and maximum value is a completely un-healthy state. Therefore, relatively more persistent sensor readings within the non-allowed ranges have more influence on the final outcome of the computed health signal value. Therefore this characteristic decreases the amount of potential fault-negative situations and aims for appropriate health signal generation by RADE.

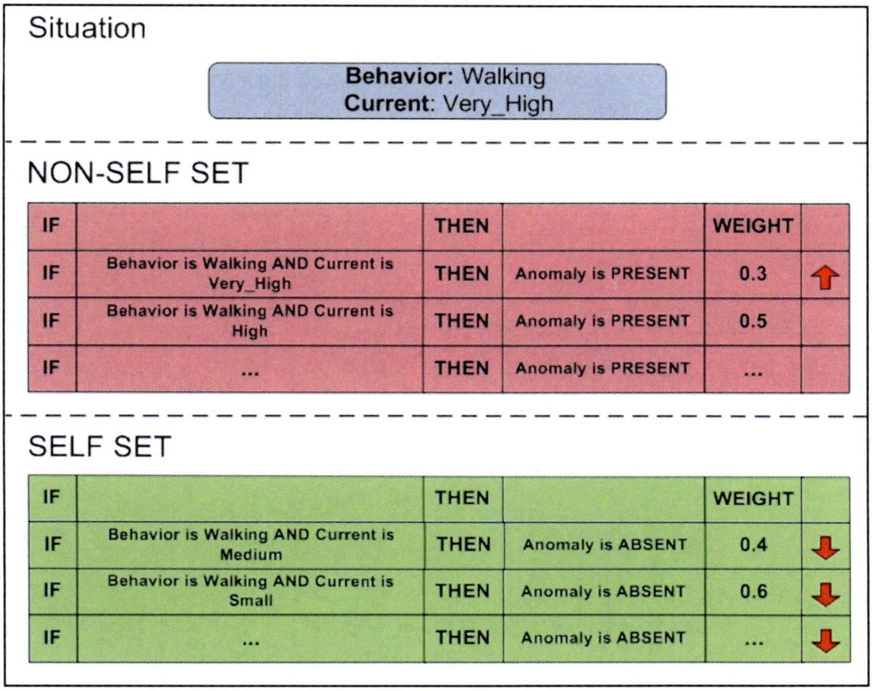

Situation

Behavior: Walking
Current: Very_High

NON-SELF SET

IF		THEN		WEIGHT	
IF	Behavior is Walking AND Current is Very_High	THEN	Anomaly is PRESENT	0.3	↑
IF	Behavior is Walking AND Current is High	THEN	Anomaly is PRESENT	0.5	
IF	...	THEN	Anomaly is PRESENT	...	

SELF SET

IF		THEN		WEIGHT	
IF	Behavior is Walking AND Current is Medium	THEN	Anomaly is ABSENT	0.4	↓
IF	Behavior is Walking AND Current is Small	THEN	Anomaly is ABSENT	0.6	↓
IF	...	THEN	Anomaly is ABSENT	...	↓

Fig. 7.6 Functioning principle of RADE and the dynamically changing weights within the self and non-self rule sets (servo position is not evaluated here).

7.4 Experiments Done with AIS Inspired RADE and Results from Experiments

7.4.1 Test-Bed Setup for RADE Approach

In order to practically demonstrate the usefulness of the RADE method, several experiments have been done with the hexapod robot platform OSCAR-2 (introduced in Chapter 3.2.2.2). The robot OSCAR-2 has been selected for the experiments with RADE since it has modified legs (Chapter 3.2.2.2, Figure 3.6, Figure 3.7), which allow for simulated leg malfunctions. By removing some pins, which replace some of the screws in the leg, the leg segment can get "artificially" malfunctioned, therefore the hexapod robot can simulate a faulty situation.

The current measurements from the 18 modified analogue servos and acceleration values for the robot OSCAR-2 (Chapter 3.2.2.2, Figure 3.6) were done with National Instruments DAQ equipment "NI SCB-68" with an additional interface and with National Instruments LabView software. Additional programming was developed in Labview for data acquisition, measurement, preprocessing, and data logging of the current signals from the 18 robot servos and acceleration values of the robot during the experiments. The pre-processing of the data was related to reducing the noise in the measurements, averaging of the values, etc. The pre-processed logged data was then fed into the RADE approach for faulty/anomaly condition detection.

The test-bed setup is schematically represented in Figure 7.7

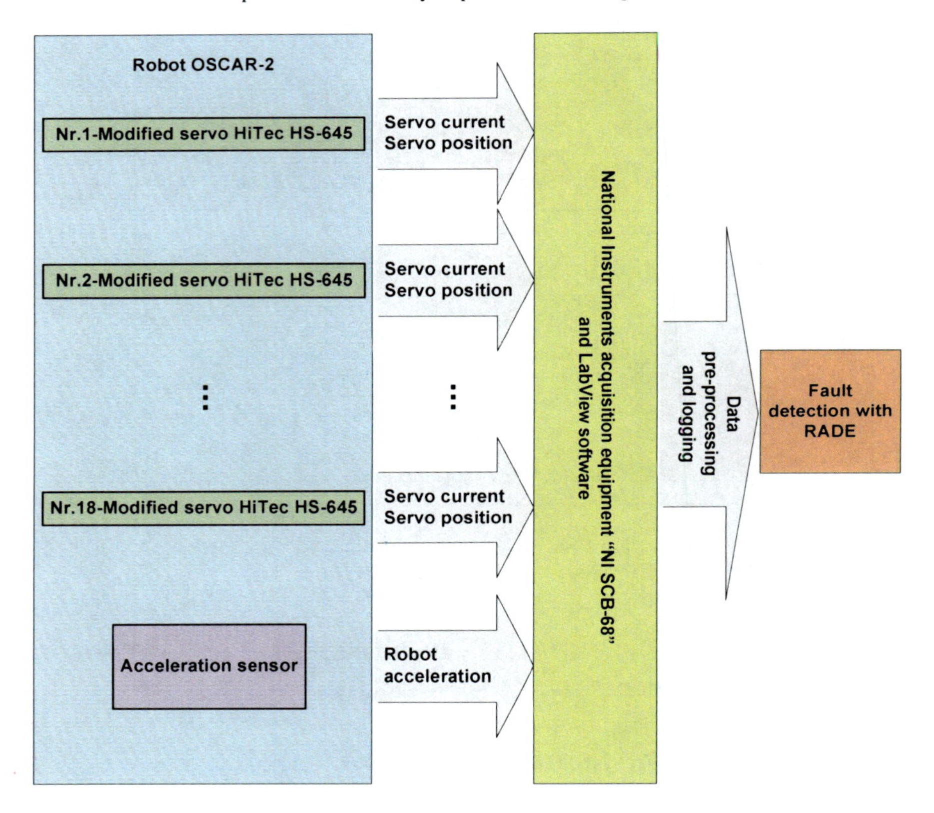

Fig. 7.7 Schematic representation of test-bed setup with robot OSCAR-2, National Instruments hardware and LabView software and fault analysis with RADE.

7.4.2 *Self and Non-self Rule Sets by RADE*

RADE method is situated in the distributed ORCA architecture (explained in Chapter 4.4, Figure 4.8), and more precisely within the OCU unit, representing the monitoring unit, which monitors the correct behavior of the servos of the robot's leg.

For testing purposes, some fuzzy logic rules have been defined for the self and non-self sets. The fuzzy rules therefore represent situations in which the robot is functional or situations in which there may be anomalies present in the system.

Rules in the *non-self set* describing the potential anomaly situation can be defined in the following way:

> IF Current is Very_High THEN Anomaly is Present WITH weight_factor_Z;

The weight_factor_Z has initial value of 0.5 and changes over time, depending on the self / non-self circumstances.

For monitoring the anomaly for simple walking behavior (given command to walk) the following rule is used:

> IF Acceleration is Very_Low THEN Anomaly is Present WITH weight_factor_Z;

The weight_factor_Z has initial value of 0.5 and changes over the course of time, depending on the self / non-self circumstances.

This particular rule has been selected since it has been observed that the robot OSCAR-2 for walking with self-organizing emergent walking gait with distributed pressure (Chapter 5.3) decreases or stops its walking when a leg malfunctions and the leg foot sensor is not touching the ground. This situation can occur also if the leg segment which malfunctions heavily inhibits the robots ability to proceed with its walking.

Rules for the *self-set* describing the situation when the robot is functional can be generated in an automatic manner since they are <u>at least</u> represented by the complementary version of the *non-self* set, although the *self* set can also contain additional rules which may better represent the non-anomalous situation and correct robot operation. For example, the initial complementary set of rules for describing the *self-set* or correct robot functioning can be defined as follows:

> IF Current is Very_Low THEN Anomaly is Absent WITH weight_factor_Z;
> IF Current is Low THEN Anomaly is Absent WITH weight_factor_Z;
> IF Current is Medium THEN Anomaly is Absent WITH weight_factor_Z;
> IF Current is High THEN Anomaly is Absent WITH weight_factor_Z;

The weight_factor_Z has initial value of 0.5 and changes over time, depending on the self / non-self circumstances.

And in relation with the observed "Acceleration" parameter, the rules for *self-set* can be defined as follows:

> IF Acceleration is Low THEN Anomaly is Absent WITH weight_factor_Z;
> IF Acceleration is Medium THEN Anomaly is Absent WITH weight_factor_Z;
> IF Acceleration is High THEN Anomaly is Absent WITH weight_factor_Z;

IF Acceleration is Very_High THEN Anomaly is Absent WITH weight_factor_Z;

The weight_factor_Z has initial value of 0.5 and changes over time, depending on the self / non-self circumstances. Depending on the implemented behaviors, additional rules can also be added. The experiments were done with the above mentioned rules that have been used to demonstrate the practical usefulness of the proposed approach.

7.4.3 Results from Experiments Done with the RADE Approach

Several experiments with RADE were done offline on real data acquired from several distinguished test cases with our robot demonstrator.

The tests are made within the following scenarios:

1. Normal walking (Figure 7.8) without RADE dynamics (Figure 7.9) and with RADE dynamics (Figure 7.10);
2. Obstacle collisions, where a robot leg hits an object (Figure 7.11) without RADE dynamics (Figure 7.12) and with RADE dynamics (Figure 7.13);
3. Mechanical problem, where a screw (simulated with a pin) in a joint falls out (Figure 7.14) without RADE dynamics (Figure 7.15) and with RADE dynamics (Figure 7.16);
4. Servos "gamma," "beta," and "alpha" (Chapter 5.1, Figure (a)) are each turned off during several experiments, simulating a potential anomaly situation of disconnected servos:
   - Servo "gamma" gets disconnected – experiment without RADE dynamics (Figure 7.17) and with RADE dynamics (Figure 7.18);
   - Servo "beta" gets disconnected – with RADE dynamics (Figure 7.19);
   - Servo "alpha" gets disconnected – with RADE dynamics (Figure 7.20);

In the pictures shown, on the X axis we have the time units. On the Y axis we can observe:

- The normalized joint motor current with values between 0 and 3, where 0 is minimum and 3 is the maximum estimated value (in amperes);
- The normalized acceleration level with values between 0 and 1.5, where 0 is minimum and 1.5 is the maximum estimated value (in gravity acceleration units);
- The normalized anomaly level with values between 0 and 1, where 0 is minimum and 1 is the maximum level of anomaly.

Experiment: Normal walking (**Figure 7.8*****) – without RADE dynamics (*****Figure 7.8*****) and with RADE dynamics (*****Figure 7.9*****);***

Fig. 7.8 Robot OSCAR-2 during normal walking.

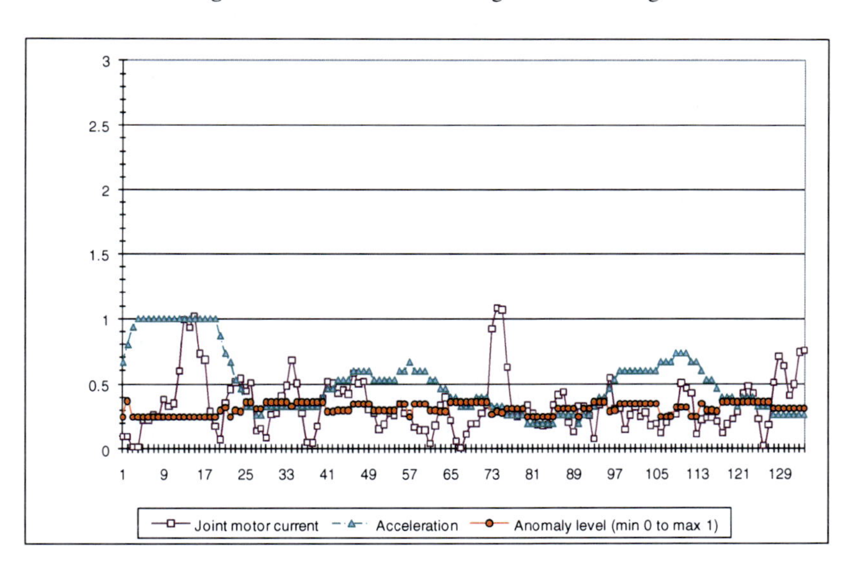

Fig. 7.9 Experiment without RADE dynamics - Measured anomaly level for normal walking.

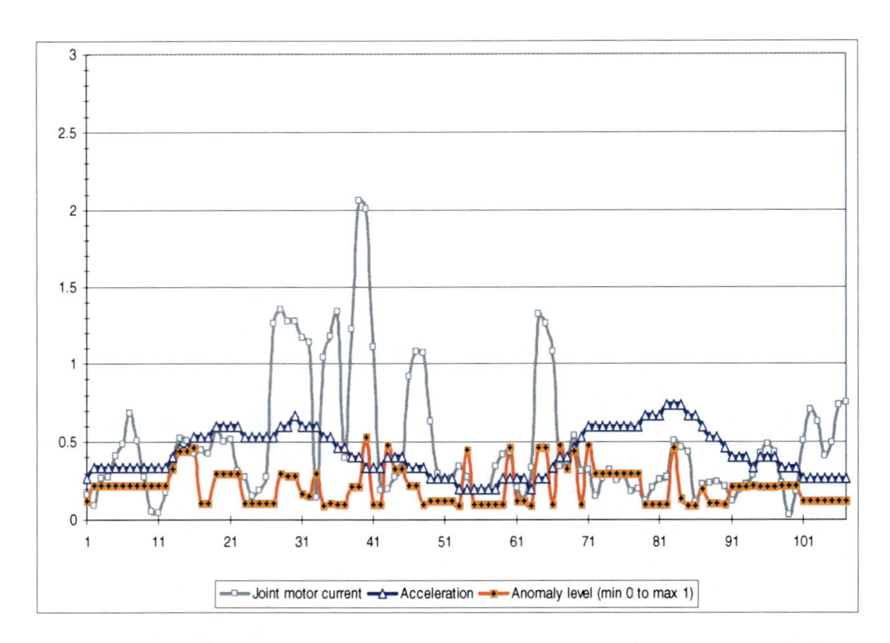

Fig. 7.10 Experiment with RADE dynamics - Measured anomaly level for normal walking.

As can be seen from the measurements presented in Figure 7.9 and Figure 7.10, and as expected, the anomaly values are under 0.5 which indicates no anomaly condition within the robot. The 0.5 value has been observed during the experiments to represent "normal" conditions.

By experiment without RADE dynamics (Figure 7.9) the weights by the fuzzy logic rules in the fuzzy rule table do not change much during the runtime. In this experiment the anomaly level stays below the 0.5 value, indicating "normal" conditions.

The anomaly level from the approach with RADE dynamics (Figure 7.10) changes dynamically depending on the immediate and relative short history values of the motor's current for the monitored servos as well as from the acceleration of the robot. However despite the weight changes from fuzzy logic rules and the dynamics within RADE, the anomaly level still stays under 0.5, which indicates a non-anomaly situation.

Experiment: Obstacle collisions, where a robot leg hits an object (Figure 7.11) – without RADE dynamics (Figure 7.12) and with RADE dynamics (Figure 7.13);

Fig. 7.11 Robot OSCAR-2 and leg collision with an object.

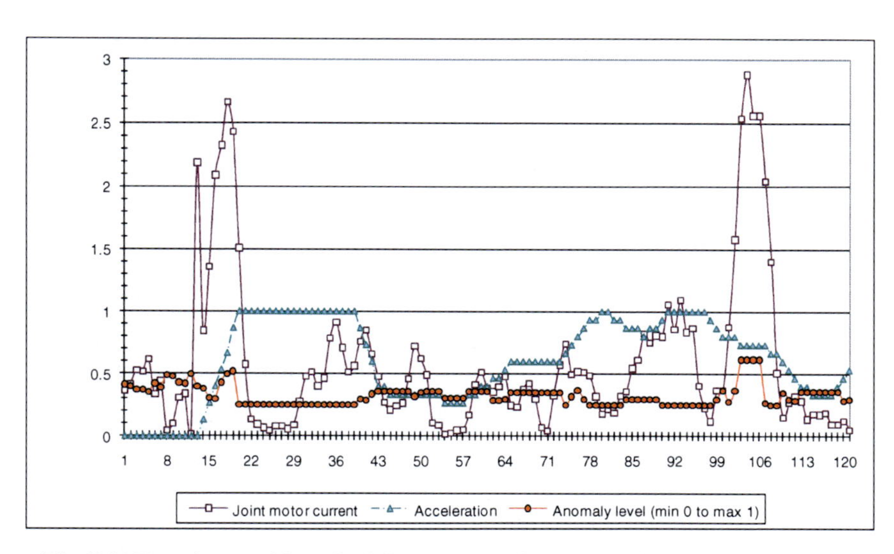

Fig. 7.12 Experiment without RADE dynamics - Obstacle collisions at 27s and 111s.

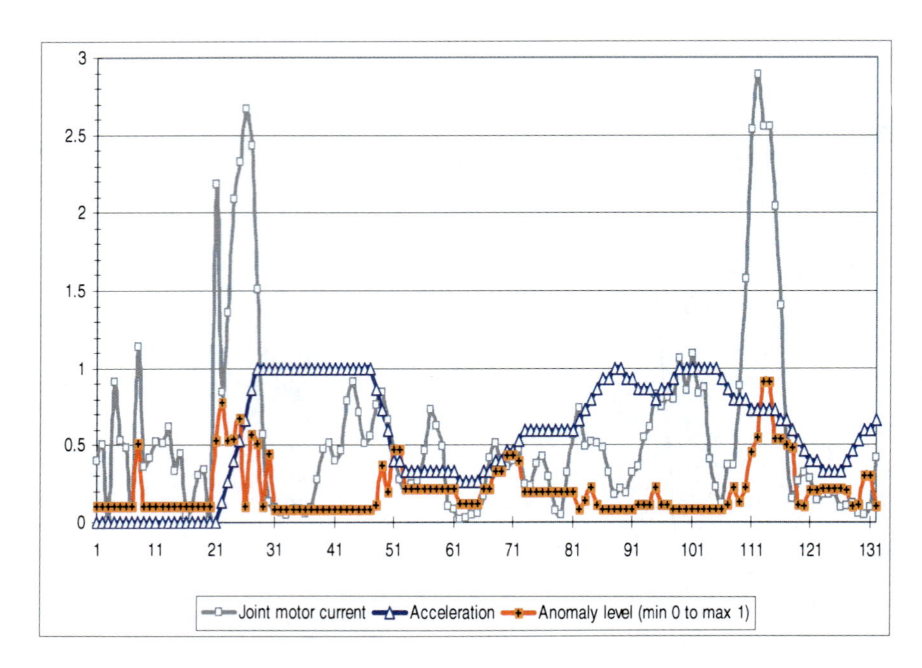

Fig. 7.13 Experiment with RADE dynamics - Obstacle collisions at 27s and 111s.

In Figure 7.11, a situation of obstacle collisions is shown, where a leg repetitively hits an obstacle on the way. By experiment without RADE dynamics (Figure 7.12) the weights from the fuzzy logic rules in the fuzzy rule table do not change during the runtime and therefore there is no adaptation to the situation but rather only evaluation of the fuzzy rules corresponding to the situation which are manually coded and not optimized for such a particular situation. Therefore the anomaly level at the first motor current peak at about 17s, indicating a leg hitting an obstacle, does not rise over 0.5 value. At the second motor current peak at about 105s, indicating a leg hitting an obstacle again, rises slightly to a value of 0.6 which does not indicate a strong anomaly situation.

By experiment with RADE dynamics, (Figure 7.13) the weights from the fuzzy logic rules change dynamically over time, depending on the immediate and relative short history values of the motor's current for the monitored servos as well as from the acceleration of the robot.

As can be seen in the Figure 7.13, the current is extremely high at two moments (27s and 111s), representing a situation where a leg is hitting an obstacle. In that case, the acceleration level is still rather normal, and anomaly level rises dynamically from short to 0.8 in the first situation and to 0.9 in the second situation. Such a high anomaly level which appears for a rather short time would further make an indication to the OCU monitoring part to take action on possibly automatically changing the pattern of walking to allow the robot to overcome the obstacle on its way. Hence, this aids in achieving the *self-organization* and *self-reconfiguration* property of the robot.

By comparing the experiments done <u>without</u> RADE dynamics and <u>with</u> RADE dynamics it can be noticed that the anomaly detection <u>with</u> RADE dynamics can easily recognize the anomaly situation in a non-fully preprogrammed way – which demonstrates the *self-adapting* characteristics of the anomaly detection system.

Experiment: Mechanical problem - a screw (simulated with a pin) in a joint falls out (Figure 7.14) – <u>without</u> RADE dynamics (Figure 7.15) and <u>with</u> RADE dynamics (Figure 7.16);

Fig. 7.14 Mechanical problem by OSCAR-2 - situation when a screw (simulated with a pin) in a joint falls out.

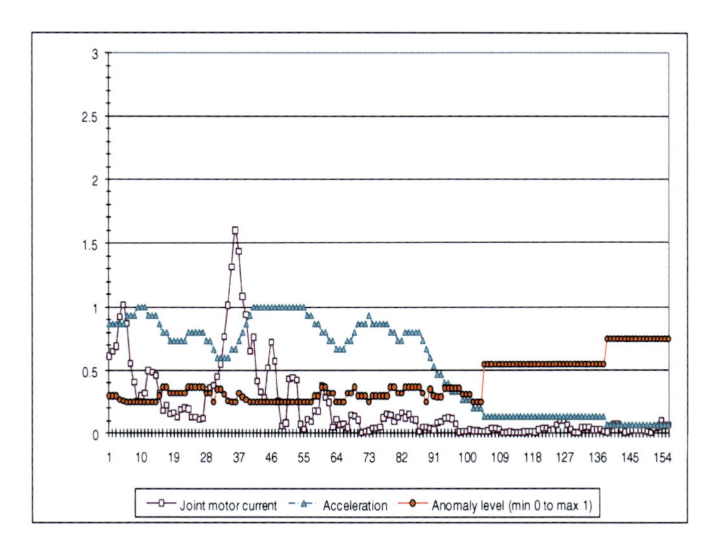

Fig. 7.15 Experiment <u>without</u> RADE dynamics - Mechanical problem - a screw (simulated with a pin) in a joint falls out at 87s.

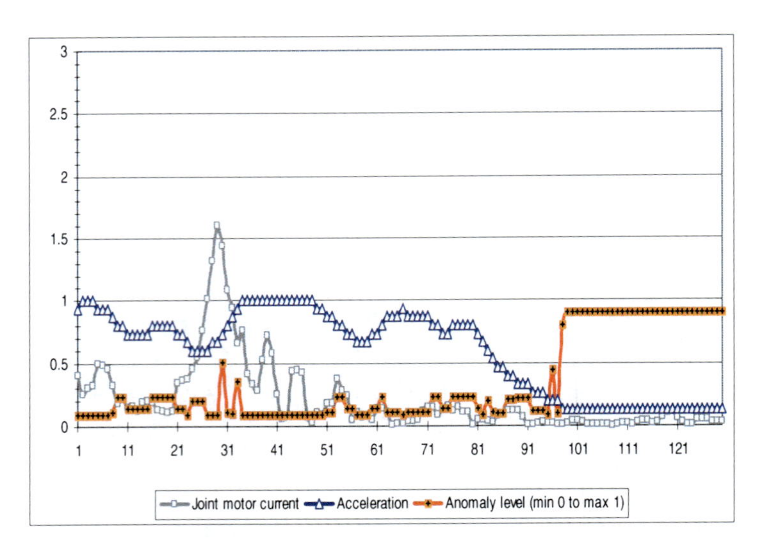

Fig. 7.16 Experiment with RADE dynamics - Mechanical problem - a screw (simulated with a pin) in a joint falls out at 80s.

In Figure 7.16, a case where a mechanical problem occurs is represented. Namely, a screw in a joint (simulated with a pin) falls out at 80s, and the joint becomes unfastened and therefore dysfunctional. This results with one additional emergent behavior of the robot – the robot stops. Since the walking behavior is executed and by this the acceleration level is monitored to not be low, the anomaly level rises.

By experiment without RADE dynamics (Figure 7.15), the weights from the fuzzy logic rules in the fuzzy rule table do not change during the runtime and there is no adaptation to the situation but rather only evaluation of the fuzzy rules corresponding to the situation which are manually coded and not optimized for such a particular situation. As can be seen from Figure 7.15, the anomaly level rises to 0.8 (due to the acceleration level dropping near 0) only at around 137s - representing very slow and unreliable reaction to the anomaly situation.

By experiment with RADE dynamics (Figure 7.16), the weights from the fuzzy logic rules change dynamically over time, depending on the immediate and relative short history values of the motor's current for the monitored servos as well as from the acceleration of the robot. The slightly dropping of acceleration in correlation with lower motor current results in a dynamic rise of the anomaly level to a value of 0.9 at about 96s. In this case, the anomaly level also persists on that level, which makes indications for the OCU to generate a signal to amputate the non-functional leg or to initiate reconfiguration of the whole robotic system (Reconfiguration of the robot is described in Chapter 8).

With such a fast reaction to an anomaly situation and with a rather distinctive anomaly level, RADE demonstrates that with such *self-adapting* characteristics it can be more reliable and better for anomaly detection than the conventional manually coded non-optimized fuzzy rule base (without RADE dynamics).

Experiments: Servos "gamma" (*Figure 7.17, Figure 7.18*), "beta" (*Figure 7.19*) and "alpha" (*Figure 7.20*) are turned off in separate experiments, simulating in that way an anomaly situation – disconnected servos:

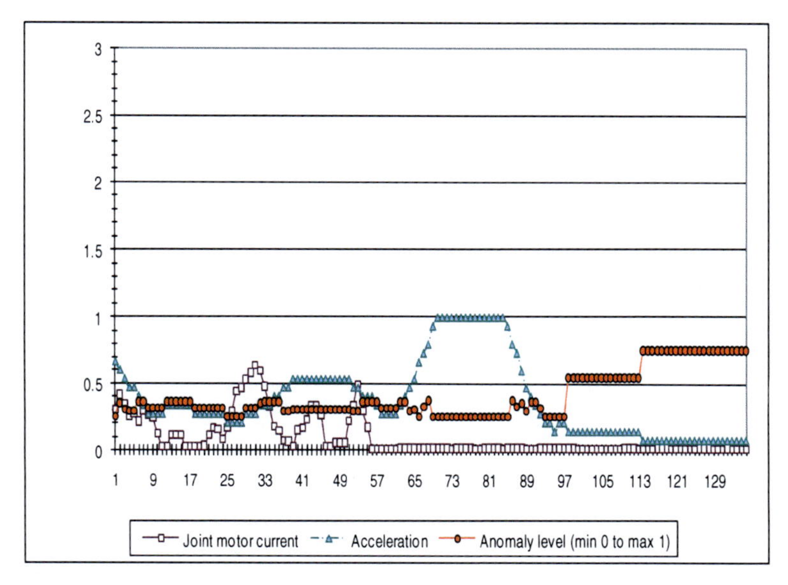

Fig. 7.17 Experiment <u>without</u> RADE dynamics - Servo "gamma" gets disconnected at time 55s.

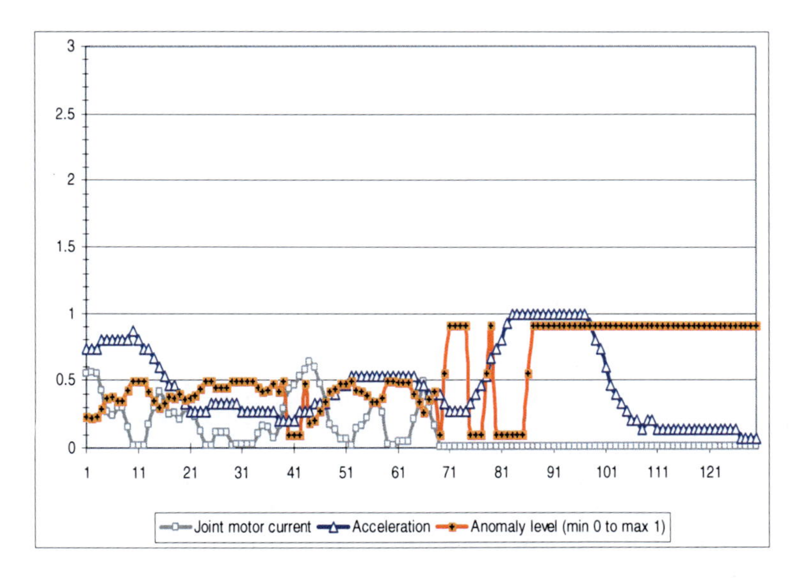

Fig. 7.18 Experiment <u>with</u> RADE dynamics - Servo "gamma" gets disconnected at time 67s.

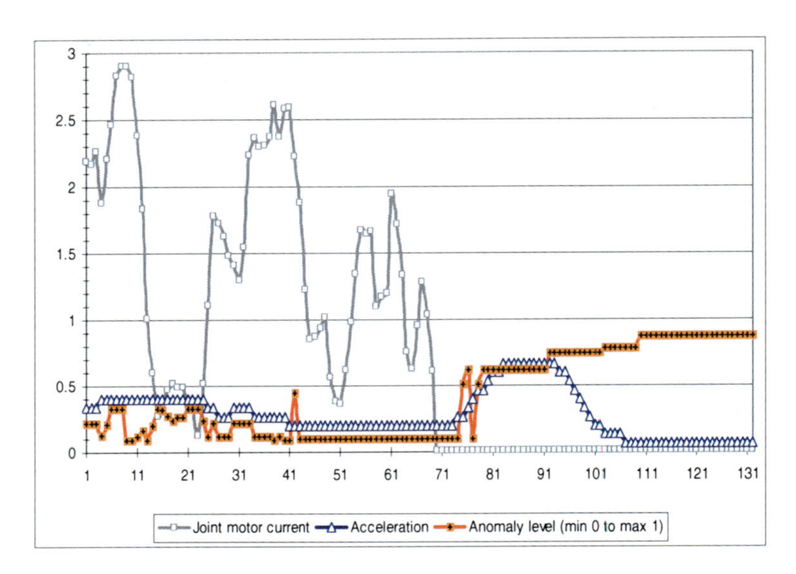

Fig. 7.19 Experiment with RADE dynamics - Servo "beta" gets disconnected at 68s.

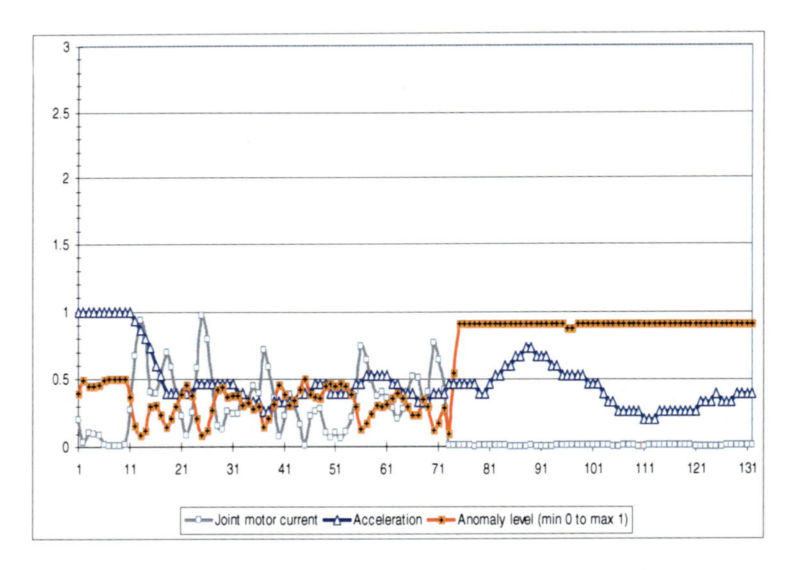

Fig. 7.20 Experiment with RADE dynamics - Servo "alpha" gets disconnected at 73s.

In Figure 7.17, results from experiment are presented where the servo "gamma" gets disconnected at 55s (simulating some failure within the servo's wires).

By experiment without RADE dynamics (Figure 7.17) the weights of the fuzzy logic rules in the fuzzy rule table do not change during the runtime and therefore there is no adaptation to the situation. Rather there is only evaluation of fuzzy rules corresponding to the situation which are manually coded and not optimized for such a particular situation. As can be seen from the Figure 7.17, the anomaly level rises to 0.8 finally at about 114s (due to the acceleration level dropping near 0 and the motor current dropping to 0) - representing rather very slow and unreliable reaction to anomaly situation.

By experiment with RADE dynamics (Figure 7.18) the weights of the fuzzy logic rules change dynamically over time, depending on the immediate and relative short history values of the motor's current for the monitored servos as well as from the acceleration of the robot. After the "gamma" servo gets disconnected at about 68s, the anomaly level rises dynamically at about 71s, and then goes down for a bit before dynamically rising again, which can be traced to the rather high acceleration level at that point in time. The anomaly level then persists at that 0.9 level from about 85s, indicating a faulty situation within the robot.

In this way the RADE approach (with its inner dynamics) again demonstrates a *self-adapting* characteristic and its usefulness for anomaly detection.

A similar situation with RADE dynamics can also be seen by experiment with servo "beta" (Figure 7.19) where after the simulated cut off servo wire at 68s, the anomaly level dynamically rises to the steady level of 0.9 at about 75s of time, which indicates severe anomaly of the robotic system. Interesting about this situation is that the servo current during this experiment is higher than the servo currents in previous experiments. Namely, this comes from the fact that due to the specific spatial locations of the legs, the "beta" motors carry more of the weight of the robot during walking and therefore consume more power, which results in higher current. However, since this is also a normal situation, the anomaly level stays rather low until an anomaly situation appears and is detected by RADE (with its inner dynamics).

Another experiment with RADE dynamics is shown in Figure 7.20, where servo "alpha" gets disconnected at 73s, which simulates cutting a wire related to servo control / power supply. The servo current therefore drops as well as the acceleration, and this initiates dynamic rise of the anomaly level to a value of 0.9 at about 75s and persist there – which indicates a faulty situation within the robot.

These 3 experiments with disconnected servos of the robot clearly represent the *self-adapting* characteristic of RADE using the Artificial Immune System paradigm and its usefulness for anomaly detection within the robot.

7.4.4 3D representation of Run-Time Dynamics by RADE Anomaly Detection Surface

During the experiments made with RADE several random snapshots are made of the dynamics within the anomaly detection system. The random snapshots do not represent some particular situation in which the robot is in, but depict how the output by fuzzy logic anomaly detection surface can be influenced from fuzzy logic rules weights change in every moment of time.

As mentioned, the weights of the fuzzy logic rules by RADE change dynamically over time, depending on the immediate and relative short history values of the motor's current for the monitored servos as well as the acceleration of the robot. This requires that the anomaly detection surface changes in the 3D domain during the RADE runtime, as shown in Figure 7.21, Figure 7.22, and Figure 7.23.

For the 3D representation in those figures, the plane axes are chosen to represent the "acceleration level" of the robot and the "alpha motor current" of one of the robot's legs, though the plane axes can represent any of the system's characteristics that need to be observed. The vertical axis in these figures represents the "anomaly level" with respect to the "acceleration level" and the "alpha motor current" of the robot.

As can be concluded from the figures Figure 7.21, Figure 7.22, and Figure 7.23, the dynamics within the RADE anomaly detection method introduces change to the anomaly detection surface and therefore guides the momentary anomaly level generation.

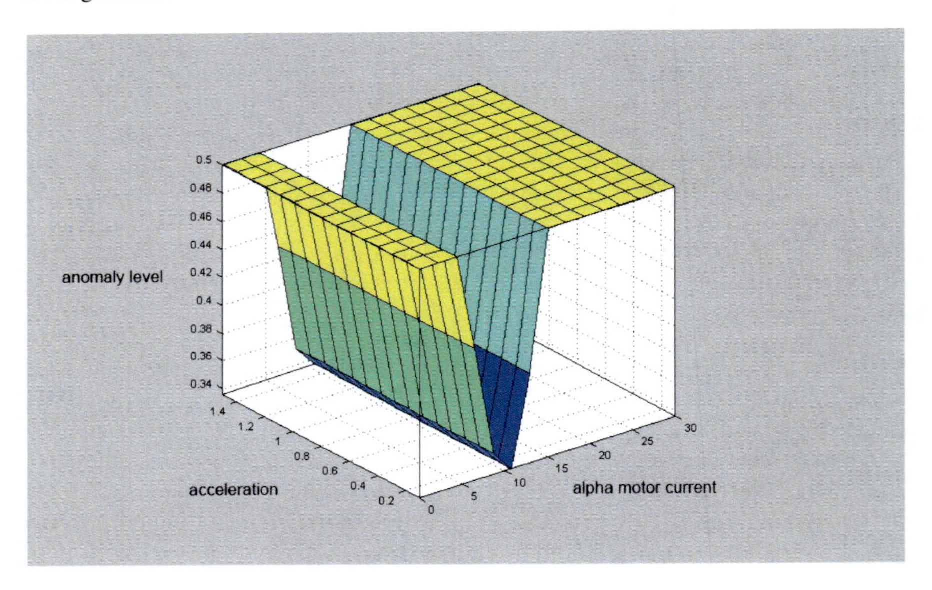

Fig. 7.21 Dynamics by RADE anomaly detection surface - snapshot 1.

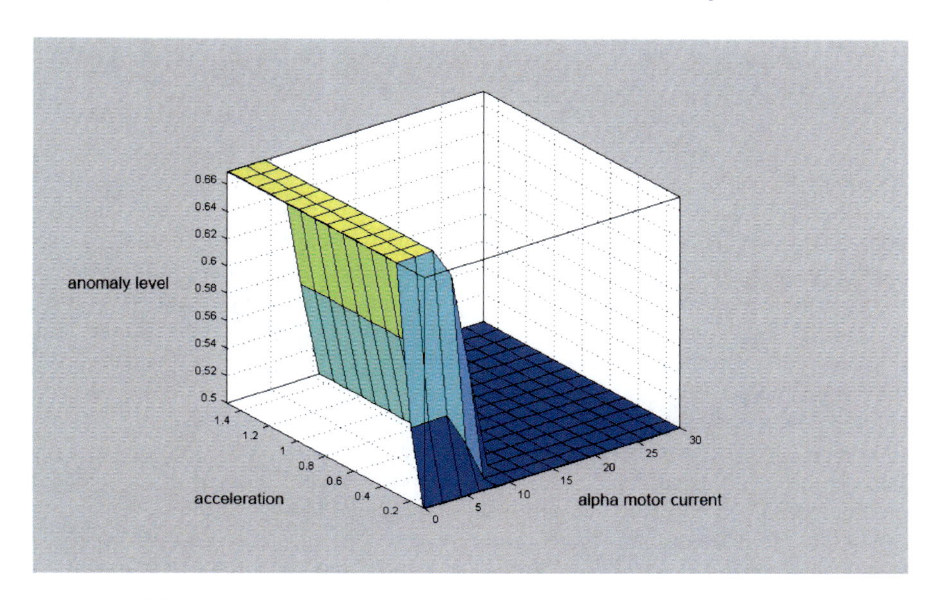

Fig. 7.22 Dynamics by RADE anomaly detection surface - snapshot 2.

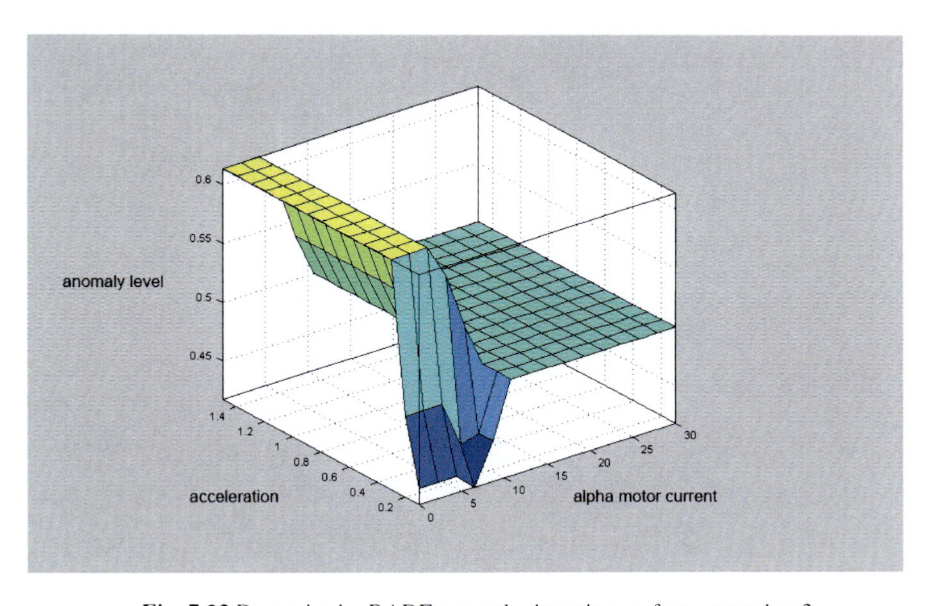

Fig. 7.23 Dynamics by RADE anomaly detection surface - snapshot 3.

7.4.5 Summary about AIS Based Anomaly Detection Approach - RADE

In this chapter an Artificial Immune System based approach named RADE (Robot Anomaly Detection Engine) for anomaly detection in walking robots has been described. The fault detection method uses the clonal proliferation as an immune system inspired mechanism to dynamically generate the level of anomaly. Based on this biological abstraction, the anomaly detection method *self-adapts* to the situation and generates an appropriate health signal which can be further interpreted. Since the method clearly avoids fully pre-programmed fuzzy logic rules for its functioning and avoids manually coding each anomaly situation that may appear, RADE considerably decreases the human effort in building such an anomaly detection system and provides reliable anomaly detection.

Further research on RADE method can envelop automatic generation of the fuzzy rules for the self and non-self sets using negative or positive selection methods with memorization. Using such automatic rules generation, the system can be able to learn online, which would mean extending the self-capability of the robot. The presented anomaly / failure detection approach RADE is useful for detection of situations where the robot has malfunctioning legs or other body parts and is therefore important for initiating reconfigurations of the robot.

Chapter 8
Approach for Robot Self-reconfiguration after Anomaly Detection within a Walking Robot System Based on Biological Inspiration - Swarm Intelligence

Fault tolerance is an important property of an autonomous robot systems. The robotic systems often have a large number of hardware and software components (control systems, sensors, actuators, power systems, communication elements) which increase the probability of and occurrence of faults within the system. And this is especially true for walking robots. These systems exhibit high degrees of freedom and have a large number of joints and actuators, which are often prone to mechanical failures. Various research studies have been done on fault detection and fault isolation for robotic systems [Zha06] [RRT07] [CMN04] [LiJ07].

In the previous chapter, a method for anomaly detection for the robotic domain has been introduced. This chapter will introduce an approach for robot reconfiguration based on biologically inspired notions which enables the robot to reconfigure itself after an anomaly has been detected within the robotic system.

For overcoming the potential robot's malfunctioning, different techniques have been considered. One of them is swarm robotics. Swarm robotics is a synonym for decentralized autonomous systems built out of many robots which communicate and cooperate with one another to accomplish some mission tasks. Due to the high number of entities in such a swarm system, swarm robots have a high potential to be considered fault tolerant and reliable systems. Namely, the failure of one or several robots may not compromise the overall system's functionality [Win06] [HLC06].

Another approach for robot reconfiguration is seen in modular robots [XMW08] [HaH06] [MuK07] [EHS06] [SMK06]. Here, the macro view from swarm robotics is shifted to a more granular view so the robots are built out of many identical functional parts - autonomous entities which when combined together and coordinated in the proper way can function like a robust robot. Due to the redundancy within the system, the failure of one or several such modules will not cause the robot to completely malfunction. The other functional modular

B. Jakimovski: Biologically Inspired Approaches, COSMOS 14, pp. 151–173.
springerlink.com

units will reconfigure and the modular robot will continue with its mission. Modular robots differ from swarm robots, since the modules in modular robots are interconnected and exploit the concept of decentralized control, where every module contributes to the overall global behavior of the system. However, it is still an ongoing research on how those separately functional elements can be combined together and coordinated autonomously to function as one robot.

In this research, a different type of granularity of the system has been considered. Namely for the walking hexapod robot, it has been assumed that the robot legs themselves act as separate entities. So it is considered that one robot leg is a functional entity built of several servos.

Therefore when taking into account this type of granularity view, there are six legs acting as six independent entities which can be considered for reconfiguration purposes.

Another contribution of this research is an idea, how to perform the reconfiguration on several such functional entities within one hexapod robot with changeable morphology.

Namely the robot reconfiguration by using inverse kinematics is often a tedious task to accomplish. Usually there are many parameters that need to be manually calibrated before every reconfiguration [AgP07]. This would be a highly limiting factor for reconfiguration within walking machines which exhibit a large number of degrees of freedom. Computing and trying every possible combination for an in-situ reconfiguration would be too computationally expensive and impractical.

The research approach presented here is named S.I.R.R. (Swarm Intelligence for Robot Reconfiguration) [JMM08] and is based on the natural phenomena of flocks of birds and schools of fish.

The S.I.R.R. concept is described in more detail after a short introduction to the swarm intelligence concept.

8.1 Overview on Swarm Intelligence – Flocking Behavior and Boids

The phenomenon of swarm intelligence, sometimes also called "collective intelligence", was introduced in the scientific work for self-organizing agents as cellular robots [BeW89], [BDT99]. Since then, swarm intelligence has been a subject of many research investigations. Reference [BDT99] defines swarm intelligence as "any attempt to design algorithms or distributed problem-solving devices inspired by the collective behavior of social insect colonies and other animal societies".

The main characteristic of swarm intelligence is the emergence (Chapter 2.5) of global behaviors as a result of local interactions with many entities, including their interaction with the environment. Such a system is decentralized without the presence of any global controller, and robust because the assigned task is completed even if some entities fail.

The swarm intelligence system is flexible and can adapt to a changing environment and local fluctuations. Several swarm based techniques such as ant colony optimization (ACO), particle swarm optimization (PSO), and boids have already been successfully applied in solving various engineering problems [DoD99] [DoS04] [KeE95] [HuE02] [Rey87] [POP07] [TSB04] [GKS04].

Flocking Behavior and Boids

Flocking and schooling are behaviors exhibited in coordinated movement of birds and fishes (Figure 8.1). Those behaviors are not a result of leadership within the group, but rather emerge from local interactions between the agents in the group.

Fig. 8.1 (a) School of fish; (b) Flock of birds.

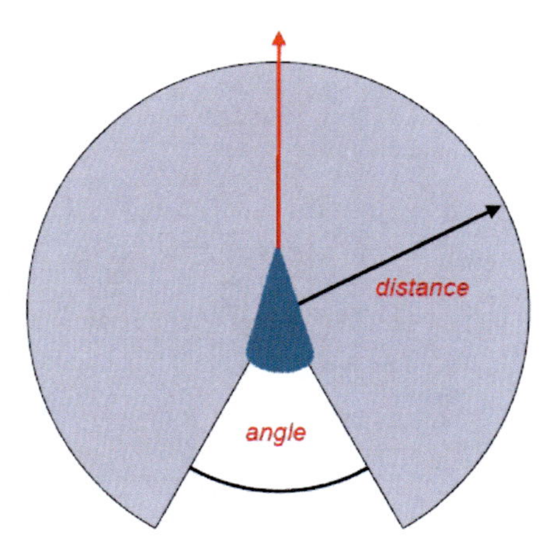

Fig. 8.2 Neighborhood of the boid.

The flocking phenomenon was observed by Craig Reynolds, who introduced boids as a distributed behavior model in to simulate the coordinated movement of a flock of birds [Rey87]. Each boid follows three rules and reacts only to boids which are within its neighborhood (Figure 8.2).

As shown on Figure 8.2 the neighborhood of the boid is represented by a radius with a center in the boid and an angle measured from the direction of the boid.

Each boid implements three general rules:

1. Separation rule: the boid tries to avoid crowding the other local boids;
2. Alignment rule: the boid moves towards the average heading of the local boids;
3. Cohesion rule: the boid moves towards the local average position of the other local boids.

These three rules are illustrated in Figure 8.3, Figure 8.4, Figure 8.5.

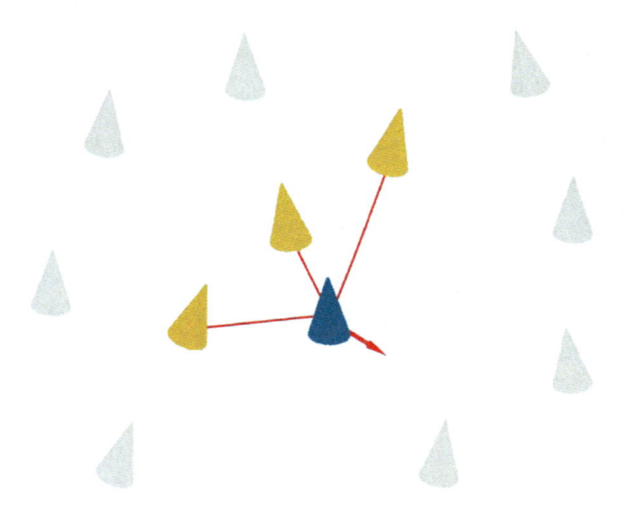

Fig. 8.3 Illustration of the separation rule.

Fig. 8.4 Illustration of the alignment rule.

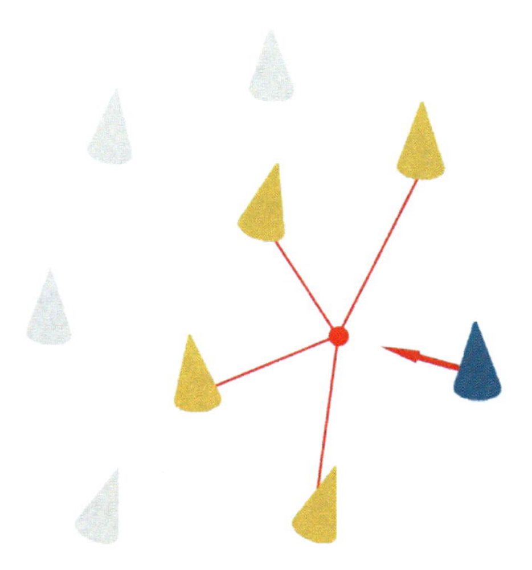

Fig. 8.5 Illustration of the cohesion rule.

Using only the three mentioned rules, emergent coordinating movement of boids can be simulated. Additional experiments have been carried out with more complex rules for obstacle avoidance [Rey88] resulting in emergent boid behavior, like splitting and reuniting after obstacle avoidance.

8.2 S.I.R.R. – Swarm Intelligence Based Approach for Robot Reconfiguration

The idea behind this concept is to introduce an approach that efficiently solves the reconfiguration problem of walking robots that have failures within their "legs". This concept is supposed to be functional for every robot with any number of legs greater than two, spatially configured in a circle. It considers shifting from the macro view seen in swarm robots, where the swarm robots colony is represented by identically built robots, to the micro view seen by the modular robots, where the modular robot is built of modules that have identical structures. Namely, with the six-legged walking robot demonstrator OSCAR (Chapter 3.2.2), the legs have an identical structure (different for various OSCAR versions, but identical for each of the hexapod robots). They are spatially configured in a circle and can be considered as boids. Since the robot has an intrinsic symmetry, with three legs on one side and three on another, the legs can be considered members of two groups of boids - Figure 8.6. The dotted line represents the line of symmetry of the robot. The ellipsoid lines represent the grouping of the legs.

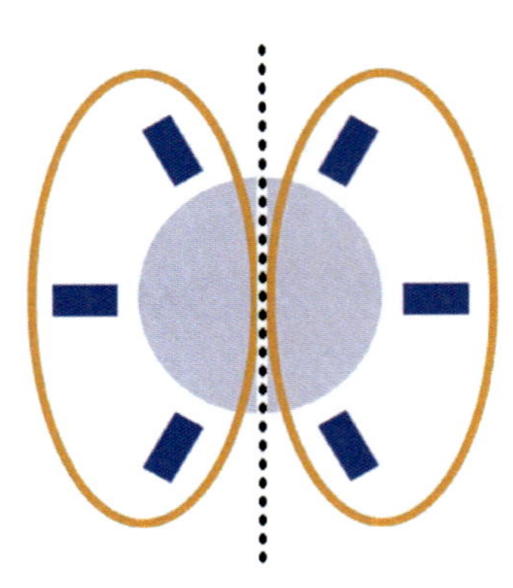

Fig. 8.6 Top schematic view of the legs of the robot forming two groups of boids; dotted line represents the line of the robot's symmetry; ellipsoid lines represent the group of boids; one rectangle represents a robot leg, which is one boid.

Each boid implements the previously explained three flocking rules, which are characteristic for boids: separation, alignment, and cohesion. In the experiments done, these rules are implemented with respect to members belonging to the same and another group of boids. Additionally within the S.I.R.R. approach, two additional rules are added to the original rule set, which contribute to even distribution of the members in the groups achieving their proper functionality after reconfiguration.

These two introduced rules can be described as follows:

1. Groups of boids are allowed to have n, n+1, or n-1 participants in comparison to the other group, where n is the number of entities (legs) in the group. In case there are n+2 or more members in one group in comparison to another, the legs which belong to the "overcrowded" group and are situated on the edge of the group (they have just one other neighbor member in the group) will be moved to another group until the "overcrowded" group's size becomes n+1.
2. Members (legs) of one group that join the other group, change their swinging and stance end-position parameters with respect to the parameters characteristic for of their new group.

The second rule is important for the leg that moves to the other side of the line of symmetry. The leg changes the swinging and stance parameters, so it swings and stans in the same manner as the other leg of that group.

Examples of the implementation of these rules is illustrated in Figure 8.7 (a), (b), (c), (d).

Figure 8.7 (a) represents the top schematic view of the robot, where the dotted line represents the robot's line of symmetry and two ellipsoid lines refer to the two groups of boids with distribution of their members (robot's legs). Figure 8.7 (b) represents an example scenario, where 2 legs belonging to one group have malfunctioned. In this case, the number of members in the left group equals one, while the number of members on the right side equals three. This means that in the right group we have n+2 members, 2 more than in the left group. Such a situation triggers the first additional rule which was already explained, and one of the two outer placed members in the right group joins the left group as shown on Figure 8.7 (c).

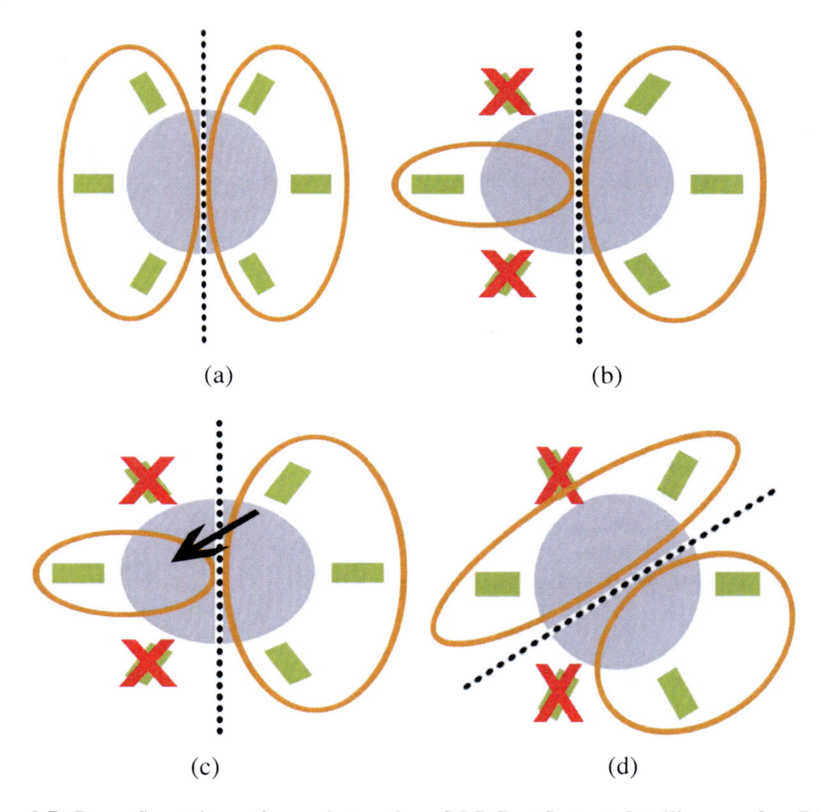

Fig. 8.7 Reconfiguration of a robot using S.I.R.R. (Swarm Intelligence for Robot Reconfiguration); (a) Top schematic view of the legs of the robot forming two groups of boids; (b) case when two legs of the robot have malfunctioned; (c) member(leg) from "crowded" group joins another less "crowded" group; (d) situation after reconfiguration. The dotted line represents the robot's line of symmetry.

In this scenario, we have chosen the upper member from the right group to join the left group, but this would be the same functionally if we instead choose the bottom member in the right group for transfer to the left one. Figure 8.7 (d) represents the situation after reconfiguration has taken place.

Due to the reconfiguration, the robot's line of symmetry has a new orientation which is represented by the dotted line. Two ellipsoid lines refer to the reconfigured groups of the boids with their redistributed members.

As a result of the second rule, the member leg that changed to the left group also assumes the new parameters for swinging and stance from the left group. Therefore, its movement is compatible with the movement of the member-leg already present in that left group.

8.2.1 Simulation of S.I.R.R Based Hexapod Robot Reconfiguration

Before testing the practical usefulness of S.I.R.R. approach on a real hexapod robot, two simulation test scenarios have been conducted in order to verify the functioning of the S.I.R.R. approach. One simulation was done without using the S.I.R.R. approach and another with using the S.I.R.R. approach.

The test scenarios consider several robot reconfigurations, after some legs on the robot have malfunctioned. For every test scenario, the robot at start is in a fully functional configuration. Then the simulation shows how some randomly chosen legs malfunction, and reconfiguration of the robot is performed.

The scenario applied in both simulations (without using the S.I.R.R. approach and with using the S.I.R.R. approach) can be described as follows:

1. leg 0 malfunctions; reconfiguration;
2. leg 2 malfunctions; reconfiguration;
3. leg 3 malfunctions; reconfiguration;
4. leg 4 malfunctions; reconfiguration.

When a leg malfunctions, the leg is first centered and then is lifted up. Then the reconfiguration takes place.

8.2.1.1 Comparative Simulation – without Using the S.I.R.R. Approach

The first simulation test was on performing the aforementioned robot reconfiguration scenario without using the S.I.R.R. approach. The results from the simulation are presented in Figure 8.8. For clearer distinction, the legs on the robot are numbered. The groups of legs are separated with a dotted line, which also represents the line of symmetry of the robot. Since there are no boid rules (separation, alignment and cohesion) implemented in this simulation experiment, once the legs have malfunctioned, the other legs do not redistribute spatially within the group – i.e. the center position of the legs stays the same. Another important note here is that there is no regrouping of the legs when their “concentration” is higher on one side of robot’s line of symmetry. This is illustrated in Figure 8.8 (c), (d). Namely, afterleg 2 has malfunctioned, there is no regrouping of the legs and the robot’s ine of symmetry remains in the same position. All these factors implicate that during such a reconfiguration of the robot without the S.I.R.R. approach, the robot attains an unstable configuration due to uneven spatial distribution of the legs. This will result in the robot not being able to continue walking and performing its mission tasks.

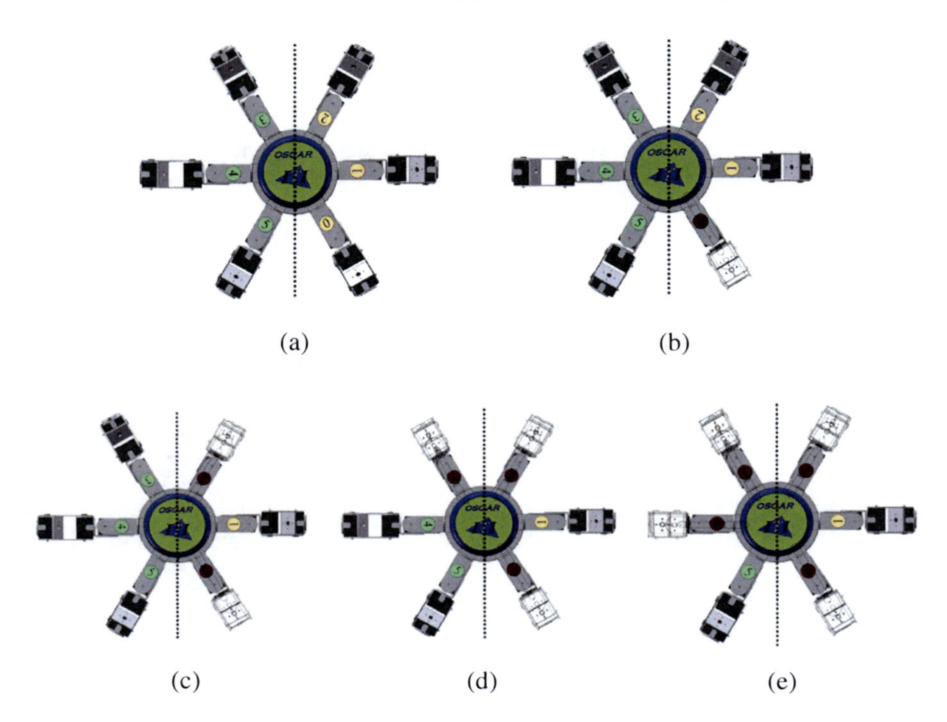

Fig. 8.8 Test scenario - reconfiguration of a robot without using the S.I.R.R. approach; (a) Top model view of fully functional hexapod robot; (b) leg number 0 has malfunctioned; (c) leg number 2 has malfunctioned; (d) leg number 3 has malfunctioned; (e) leg number 4 has malfunctioned; the dotted line represents the robot's line of symmetry.

As a direct comparison to the results of this simulation experiment, another simulation experiment for robot reconfigurations has been done, this time with the S.I.R.R. approach.

8.2.1.2 Comparative Simulation – with Using the S.I.R.R. Approach

An additional robot reconfiguration simulation experiment has been done considering the same scenario i.e. the same order by which the robot legs become malfunctional. However, in this second simulation experiment robot re-configuration has been done with using the S.I.R.R. approach. The procedure is similar to the first case – the reconfiguration takes part when some legs have malfunctioned. The results of the simulated reconfiguration including the intermediate steps are represented in Figure 8.9.

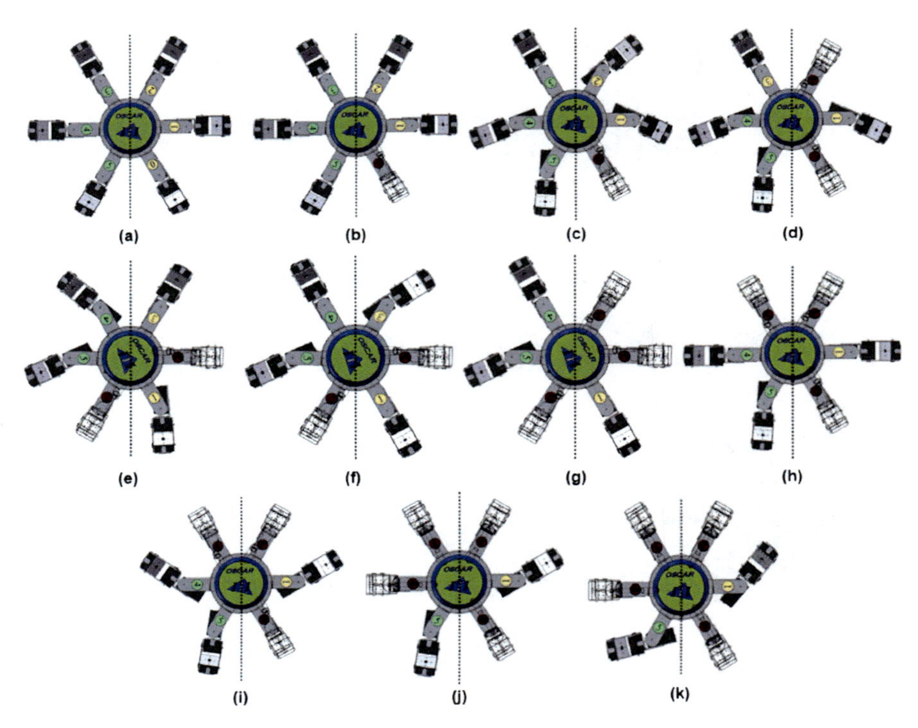

Fig. 8.9 Test scenario - reconfiguration of a robot using the S.I.R.R. approach; (a) Top model view of fully functional hexapod robot; (b) leg number 0 has malfunctioned; (c) after S.I.R.R. reconfiguration; (d) leg number 2 has malfunctioned; (e) S.I.R.R. reconfiguration with change of the symmetry axis; (f) after S.I.R.R. reconfiguration; (g) leg number 3 has malfunctioned; (h) S.I.R.R. reconfiguration with a change of the symmetry axis; (i) after S.I.R.R. reconfiguration; (j) leg number 4 has malfunctioned; (k) after S.I.R.R. reconfiguration; the dotted line represents the robot's line of symmetry.

The test scenario starts with a fully functional hexapod robot, Figure 8.9 (a). Then leg 0 of the robot malfunctions - Figure 8.9 (b). After this, the S.I.R.R. reconfiguration is done and the spatial reconfiguration of the robot's legs is performed. Figure 8.9 (c) represents the robot's spatial configuration of the legs after the S.I.R.R. reconfiguration was done. After this, leg number 2 malfunctions, Figure 8.9 (d). S.I.R.R. reconfiguration is performed, and the direction of the axis of symmetry of the robot is adjusted, Figure 8.9 (e). The result of the S.I.R.R. reconfiguration and new spatial arrangement of the legs is represented in Figure 8.9 (f). In Figure 8.9 (g), leg 3 malfunctions, S.I.R.R. reconfiguration is performed, and the direction of the axis of symmetry of the robot is adjusted, Figure 8.9 (h). Reconfigured robot using S.I.R.R. is shown in Figure 8.9 (i). In Figure 8.9 (j), leg 4 malfunctions. The final result after robot reconfiguration using S.I.R.R. is shown in Figure 8.9 (k).

In comparison with the previously simulated robot reconfiguration experiment without using the S.I.R.R. approach, the results from the reconfiguration experiment with using the S.I.R.R. approach show a better spatial reconfiguration of the robot's legs, so the robot is stable even after some legs have malfunctioned, which is not the case by the reconfiguration experiment without using the S.I.R.R. approach.

8.3 Results from Robot Reconfiguration Experiments Done with S.I.R.R. Approach on the Hexapod Robot OSCAR-2

Additionally to the simulated reconfiguration, a practical application of the S.I.R.R. reconfiguration approach has been first tried out on the hexapod robot demonstrator OSCAR-2 (Chapter 3.2.2.2) without any real leg amputations. In this experiment, the legs get malfunctioned in the same order as in the previously explained scenario for the simulation tests (so the simulation results can be compared with the real reconfiguration on hexapod robot demonstrator OSCAR-2 without any real leg amputations):

1. leg 0 has malfunctioned; reconfiguration;
2. leg 2 has malfunctioned; reconfiguration;
3. leg 3 has malfunctioned; reconfiguration;
4. leg 4 has malfunctioned; reconfiguration.

The leg malfunctions are virtual and not actually present on the robot. They are just chosen in this experiment to represent real case leg malfunctions. When a leg is marked as malfunctioned, the leg is firstly centered and then lifted up.
Then the reconfiguration using S.I.R.R. approach takes place. Results of this experiment are represented in Figure 8.10.

The experiment starts with a fully functional hexapod robot OSCAR-2, as in Figure 8.10 (a). Then leg 0 malfunctions and the S.I.R.R. reconfiguration is performed - Figure 8.10 (b). After that, leg 2 malfunctions and the S.I.R.R. reconfiguration is done - Figure 8.10 (c). Finally leg 3 malfunctions and the S.I.R.R. reconfiguration is done - Figure 8.10 (d). The only difference from the previous experiment is that in this scenario three legs malfunction instead of four, since the construction of the legs and the power of the servos of OSCAR-2 are not suitable for performing reconfiguration when only two legs are functional.

As can be seen from the experiment in the Figure 8.10 (d), the S.I.R.R. reconfiguration allows the robot to spatially reconfigure the center of the movement of its legs in such a way that even after some legs have malfunctioned, the robot can still be stable on the rest of the functional legs and be ready to continue with its mission.

The results from this experiment (Figure 8.10) can be directly compared with the results from the simulation experiment done with the S.I.R.R. approach (Figure 8.9). From both figures it can be concluded that the experiment results are identical with the simulation results, as expected.

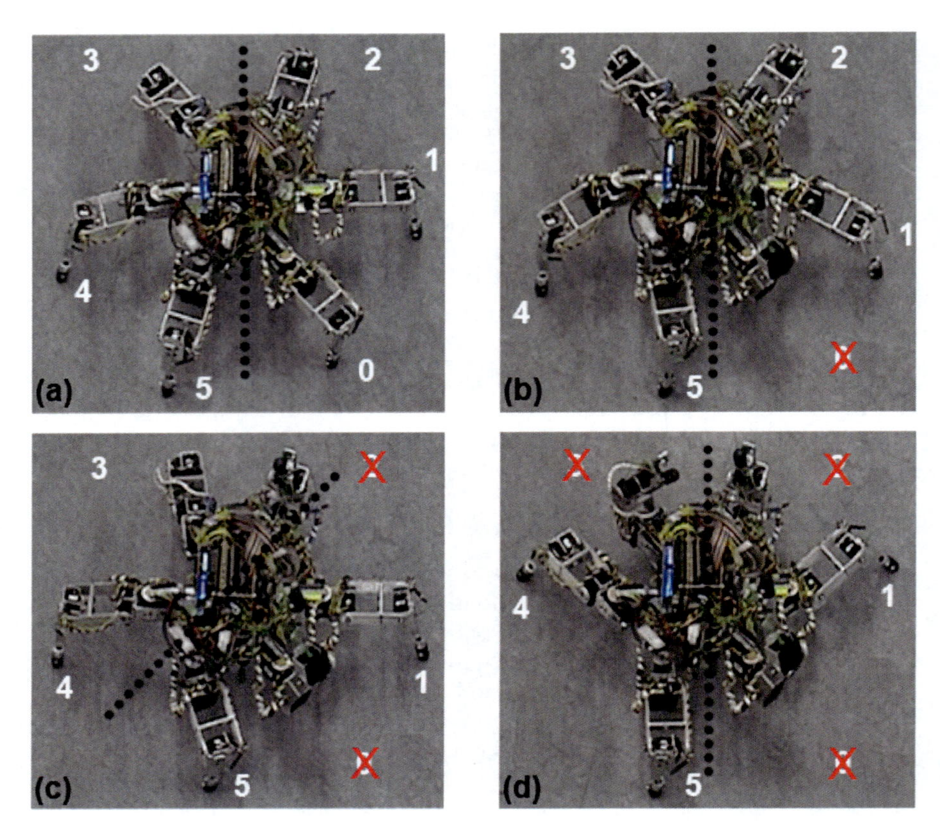

Fig. 8.10 Test scenario - reconfiguration using the S.I.R.R. approach by the hexapod robot demonstrator OSCAR; (a) Top view of the fully functional hexapod robot OSCAR; (b) leg 0 has malfunctioned and S.I.R.R. reconfiguration is done; (c) leg 2 has malfunctioned and S.I.R.R. reconfiguration is done; (e) leg 3 has malfunctioned and S.I.R.R. reconfiguration is done; the dotted line represents the robot's line of symmetry.

8.4 Results from Real Robot Reconfiguration Experiments Done with S.I.R.R. Approach and Leg Amputations on the Robot OSCAR-X

The goal of the next real robot reconfiguration experiments was to perform in-situ real time hexapod robot reconfiguration with leg amputations (done with the R-LEGAM mechanism) and enable the hexapod robot to continue with its walking despite the malfunctioning legs. This experiment was done on the hexapod OSCAR-X (Chapter 3.2.2.4).

A previously introduced innovative robot leg amputation mechanism R-LEGAM (Chapter 3.2.2.4.1) was used for doing the experiments with robot leg amputations. R-LEGAM enables the robot to amputate the malfunctioned legs on demand [JMM09].

For example, when the monitoring unit in the robot's control architecture (Chapter 4.4) detects there is an anomaly present within the leg, it sends a control signal to the ejection mechanism R-LEGAM located on the robot's leg to initiate a leg ejection. This entails amputation of the malfunctioning leg and then reconfiguration and spatial positioning of the robots legs.

The leg malfunctions are virtual and not actually present on the robot. They are just chosen in this experiment to represent real case leg malfunctions. When a leg is marked as malfunctioned, it gets amputated. Then the reconfiguration using S.I.R.R. approach takes place and the legs are reconfigured in a new spatial configuration and the robot continues with its walking.

The following demonstration scenario with following simulation of leg defects has been realized:

1. First, leg 3 becomes malfunctioned and the robot performs SIRR reconfiguration;
2. Second, leg 1 becomes malfunctioned and the robot performs SIRR reconfiguration;
3. Third, leg 5 becomes malfunctioned and the robot performs SIRR reconfiguration;

This is represented in Figure 8.11 (a) - (l).

As can be seen in Figure 8.11 (a), the robot starts with the initial six leg configuration. In the first fault case, leg 3 becomes malfunctioned and the robot control architecture sends a signal to the leg amputation mechanism to amputate the leg 3. This is shown in Figure 8.11 (b).

After that the robot performs self-reconfiguration using the S.I.R.R. approach as shown in Figure 8.11 (c) and continues with its mission.

In the second fault case, leg 1 malfunctions (Figure 8.11 (d)) and gets amputated (Figure 8.11 (e)). After that, the robot performs self-reconfiguration (Figure 8.11 (f)) and continues with walking. In Figure 8.11 (g) the third leg becomes malfunctioned and gets amputated (Figure 8.11 (h)). After that the robot performs self-reconfiguration using the S.I.R.R. approach (Figure 8.11 (i)) and continues with walking (Figure 8.11 (j) - (l)).

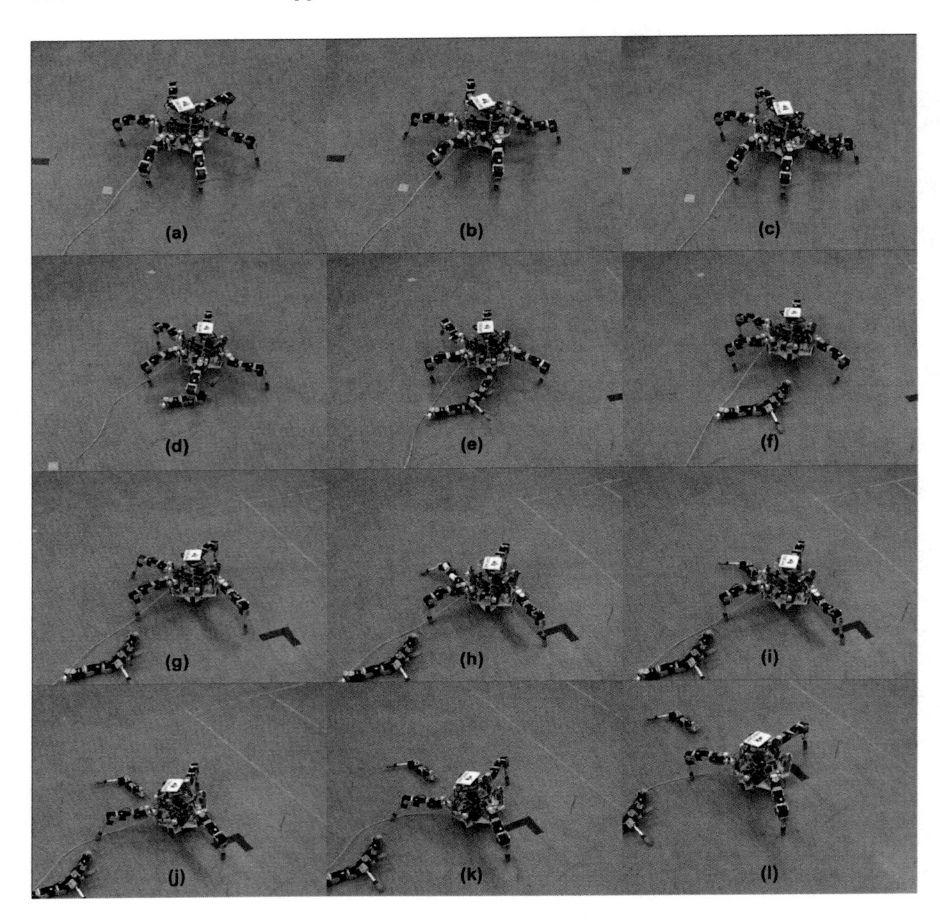

Fig. 8.11 Runtime reconfiguration of a hexapod robot OSCAR from 6 to 3 legs: (a) normal six legged configuration; (b) leg 3 malfunctions and gets amputated; (c) robot performs reconfiguration using the SIRR approach and continues walking; (d) leg 1 malfunctions; (e) leg 1 gets amputated; (f) robot performs reconfiguration using the SIRR approach and continues walking; (g) leg 5 malfunctions; (h) leg 5 gets amputated; (i) the robot performs reconfiguration using the SIRR approach and continues with walking; (j)-(l) robot OSCAR continues with its mission despite the loss of 3 legs.

8.4.1 Ground Contacts of Robot Legs for Normal Walking and for Walking with Leg Amputations and Robot Self-reconfiguration

Additional monitoring during the self-reconfiguration experiments has been done on ground contacts of the robot's legs during normal walking and walking with robot leg amputations/self-reconfiguration. The robot walking during these experiments was done using emergent walking. The results of these analyses are presented in Figure 8.12 and Figure 8.13.

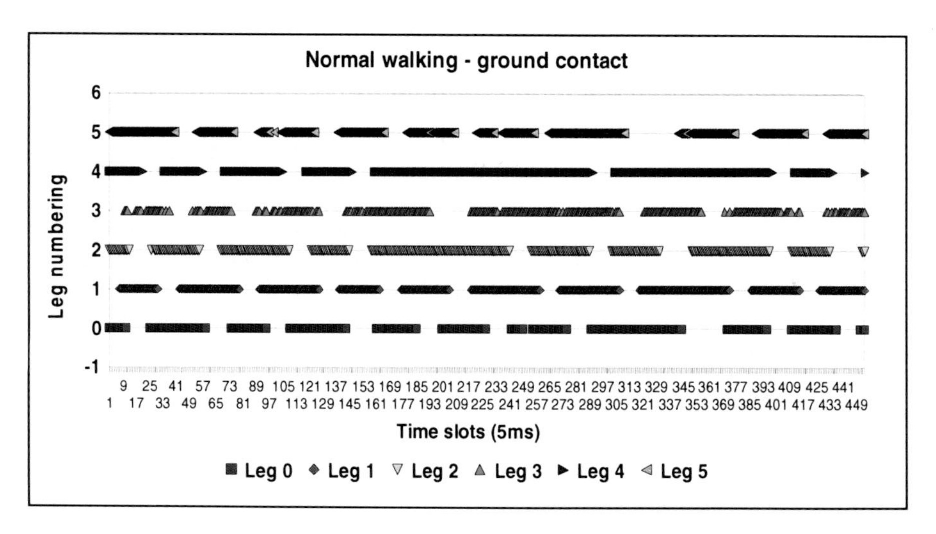

Fig. 8.12 Ground contacts of the robot's feet during normal walking of the hexapod robot.

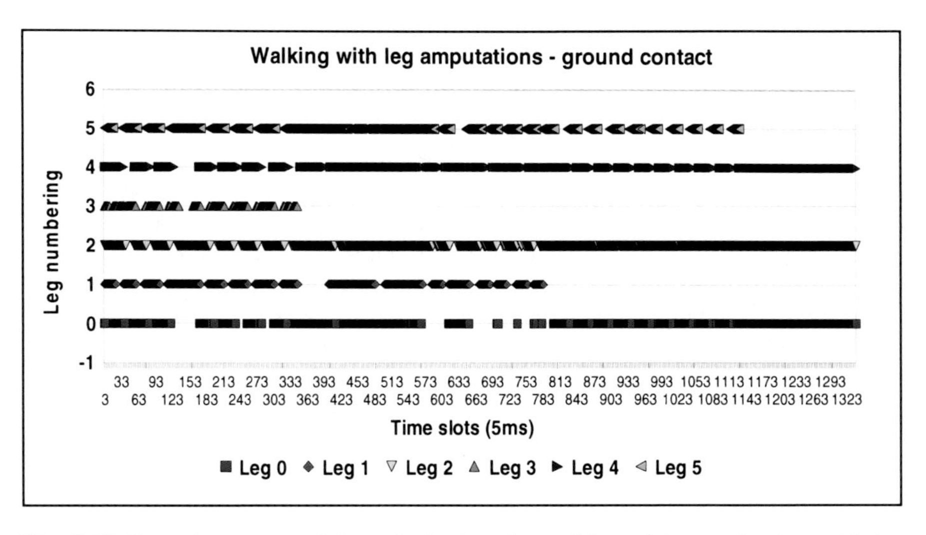

Fig. 8.13 Ground contacts of the robot's feet for walking of hexapod robot with leg amputations and self-reconfigurations at: about 363 slot of time (1st reconfiguration); about 783 slot of time (2nd reconfiguration); about 1143 slot of time (3rd reconfiguration).

The robot in these experiments is walking with a biologically inspired emergent gait (described in Chapter 5.3, [EML06]), which means the gait is not "hard-wired" or by any means predefined. By this walking, the gait pattern emerges from the local swing and stance phases of the robot's legs "joining" the group of boids at the particular robot's side after the reconfiguration has been performed.

In Figure 8.12 the chart represents the leg ground contacts for normal walking for a fully functional robot. In Figure 8.13 the chart represents the leg ground contacts of the robot walking with leg amputations where it can be seen how the legs get amputated during the experiment, the leg ground contacts are lost and the robot still continues with its walking. Leg 3 gets amputated at time slot 335; leg 1 gets amputated at time slot 785; and leg 5 gets amputated at time slot 1140. The swing phases are drastically shortened with each reconfiguration and after the time slot 1140 the robot still continues to walk, though with very shortened swing phases compared to the relatively longer stance phases.

8.4.2 Tracking of the Robot's Heading While the Robot Is Performing Self-reconfiguration with Leg Amputations

Additional measurements have been done on tracking the robot's heading while performing leg amputations and robot reconfigurations using the S.I.R.R. approach. These measurements test the straight walking and heading of the robot while it is performing leg amputations in different sequence and their influence on the robot's walking.

The solid line in figures Figure 8.15, Figure 8.17, and Figure 8.19 represents the track of the robot during its walking. The arrow lines represent the heading of the robot. The initial heading angle is 270°.

Experiment 1:

- Figure 8.14 - OSCAR-X performing leg amputations during its walking in the following order: 0, 1, 2 (from left to right and from top to bottom).
- Figure 8.15 - Tracking of the robot's heading while the robot is amputating legs during its walking in the following order: 0, 1, 2. The solid line represents the track of the robot during its walking. The arrow lines represent the heading of the robot during its walking.

Experiment 2:

- Figure 8.16 - OSCAR-X performing leg amputations during its walking in the following order: 0, 2, 4 (from left to right and from top to bottom).
- Figure 8.17 - Tracking of the robot's heading while the robot is amputating legs during its walking in the following order: 0, 2, 4. The solid line represents the track of the robot during its walking. The arrow lines represent the heading of the robot during its walking.

Experiment 3:

- Figure 8.18- OSCAR performing leg amputations during its walking in the following order: 5, 1, 2 (from left to right and from top to bottom).
- Figure 8.19 - Tracking of the robot's heading while the robot is amputating legs during its walking in the following order: 5, 1, 2. The solid line represents the track of the robot during its walking. The arrow lines represent the heading of the robot during its walking.

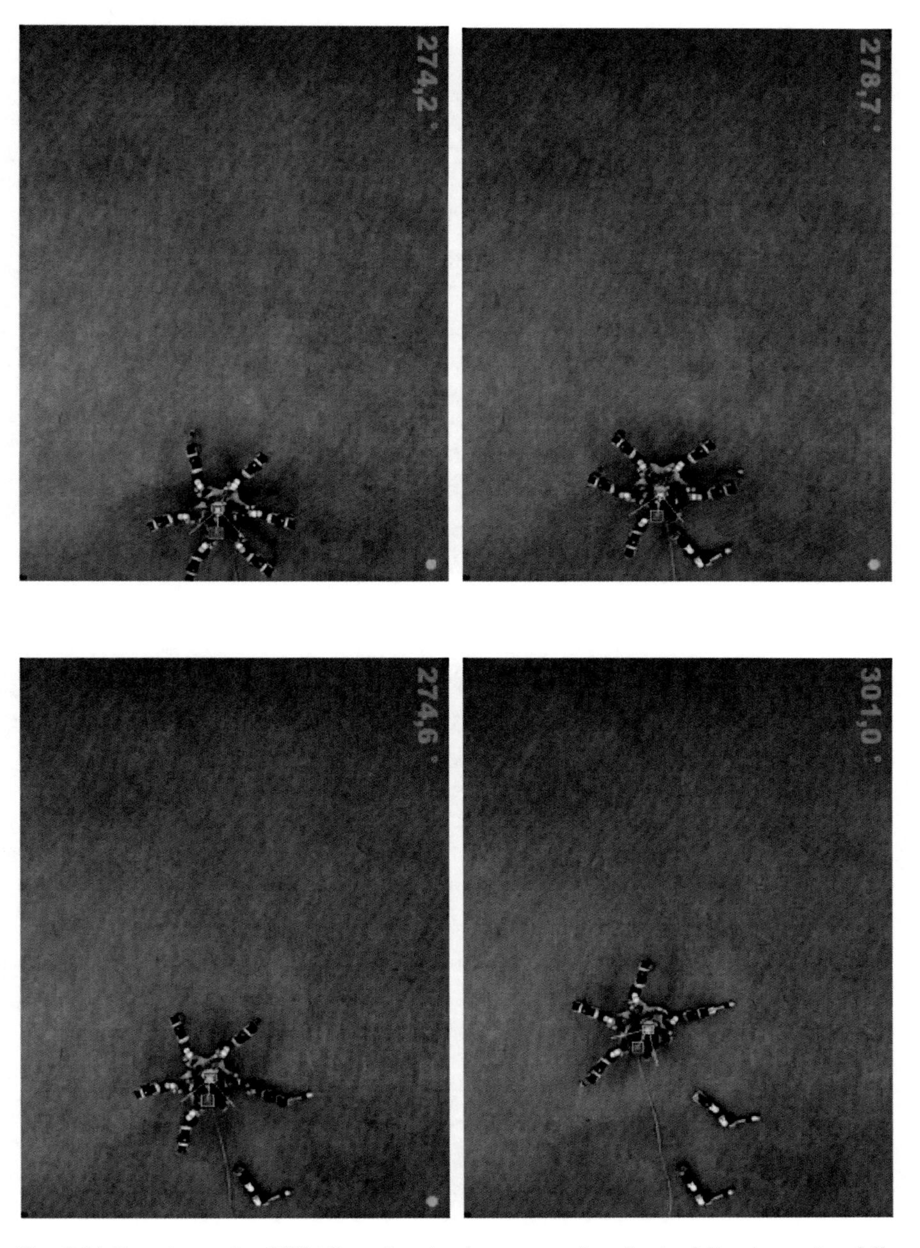

Fig. 8.14 Experiment 1 - OSCAR performing leg amputations in the following order: fully functional, leg 0 amputated, leg 1 amputated, leg 2 amputated - from left to right and from top to bottom.

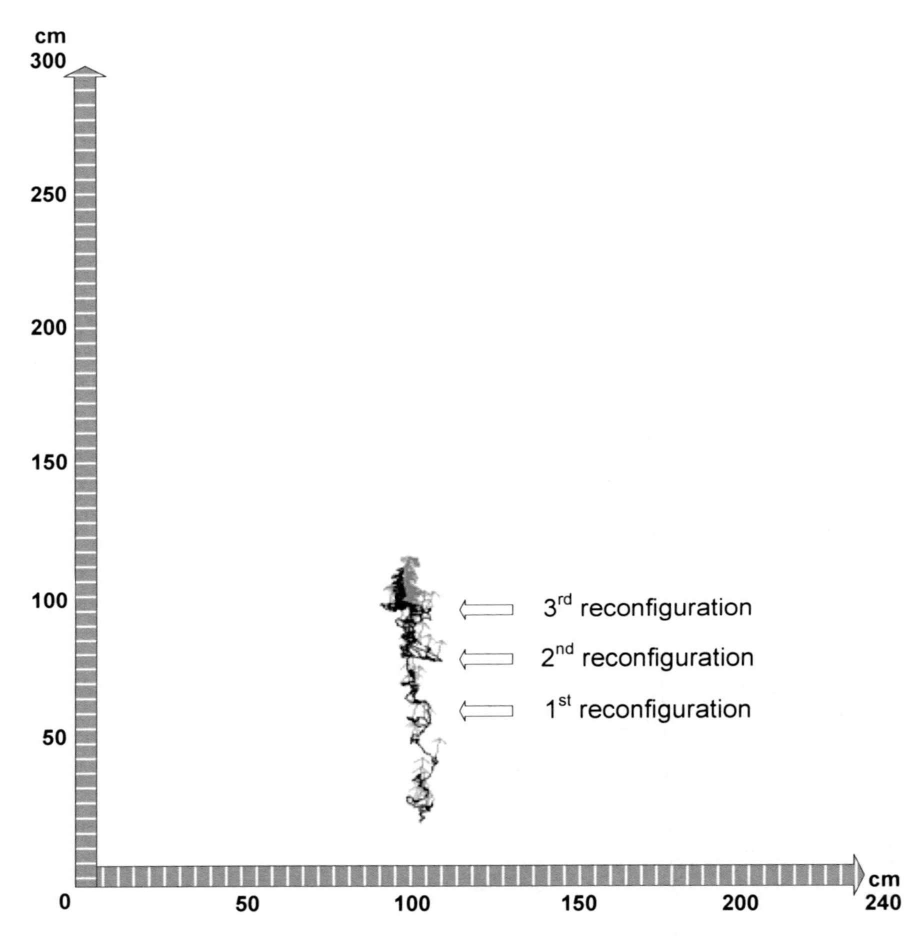

Fig. 8.15 Experiment 1 - Tracking the robot's heading while ejecting legs during its walking in the following order: 0, 1, 2.

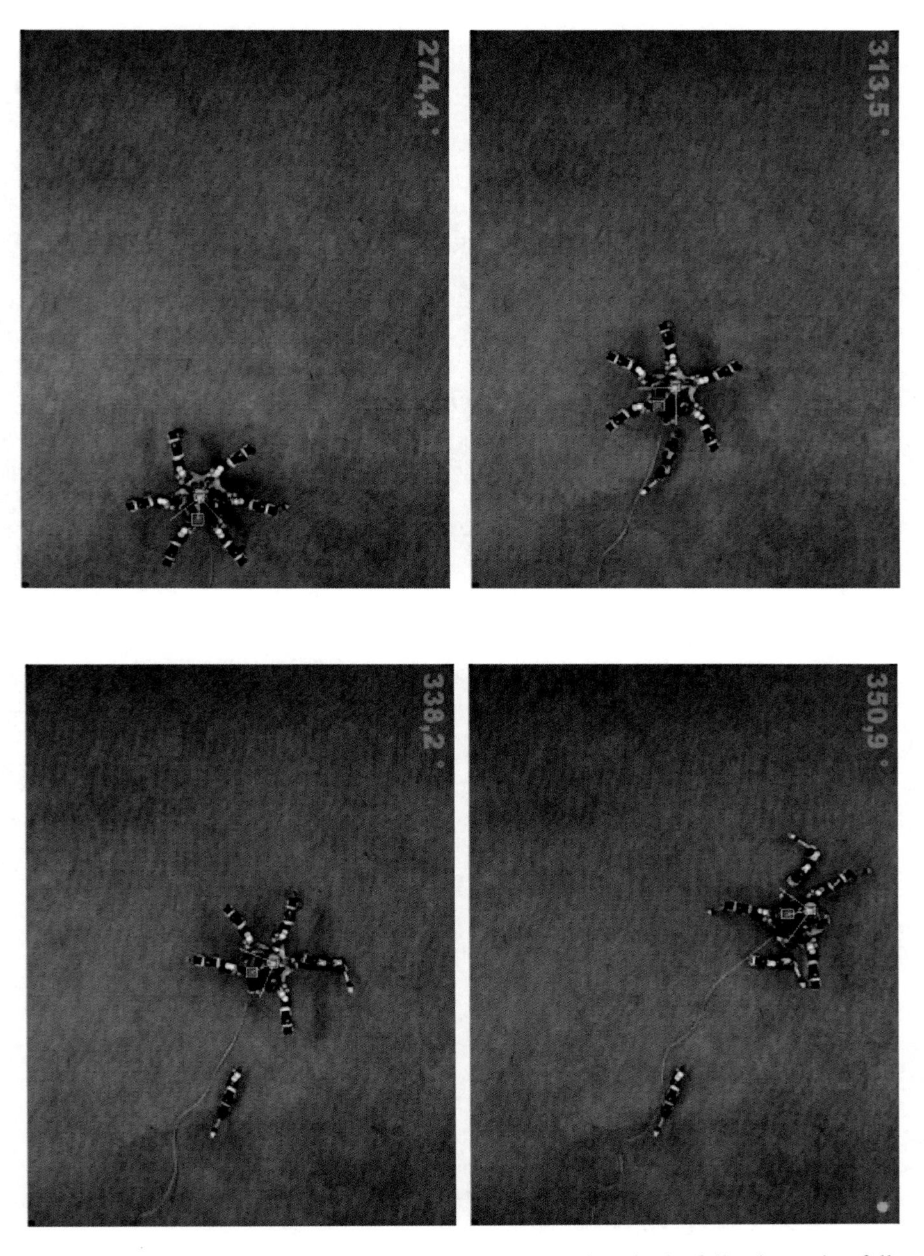

Fig. 8.16 Experiment 2 - OSCAR performing leg amputations in the following order: fully functional, leg 0 amputated, leg 2 amputated, leg 4 amputated - from left to right and from top to bottom.

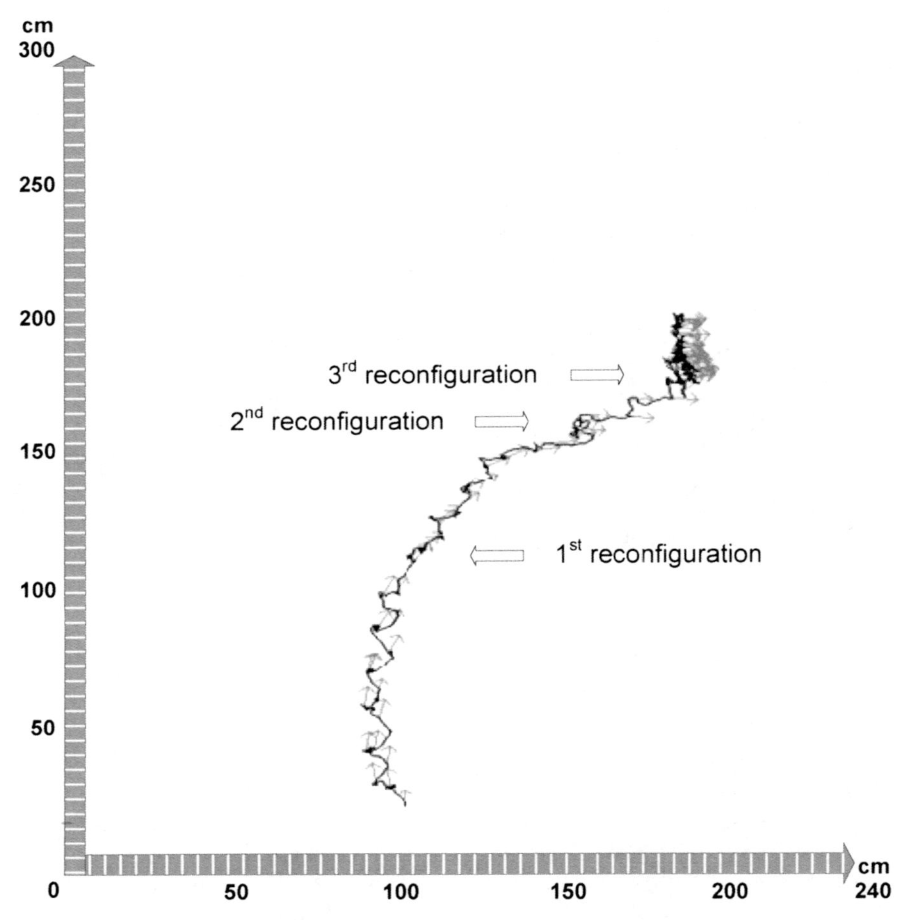

Fig. 8.17 Experiment 2 - Tracking the robot's heading while ejecting legs during its walking in the following order: 0, 2, 4.

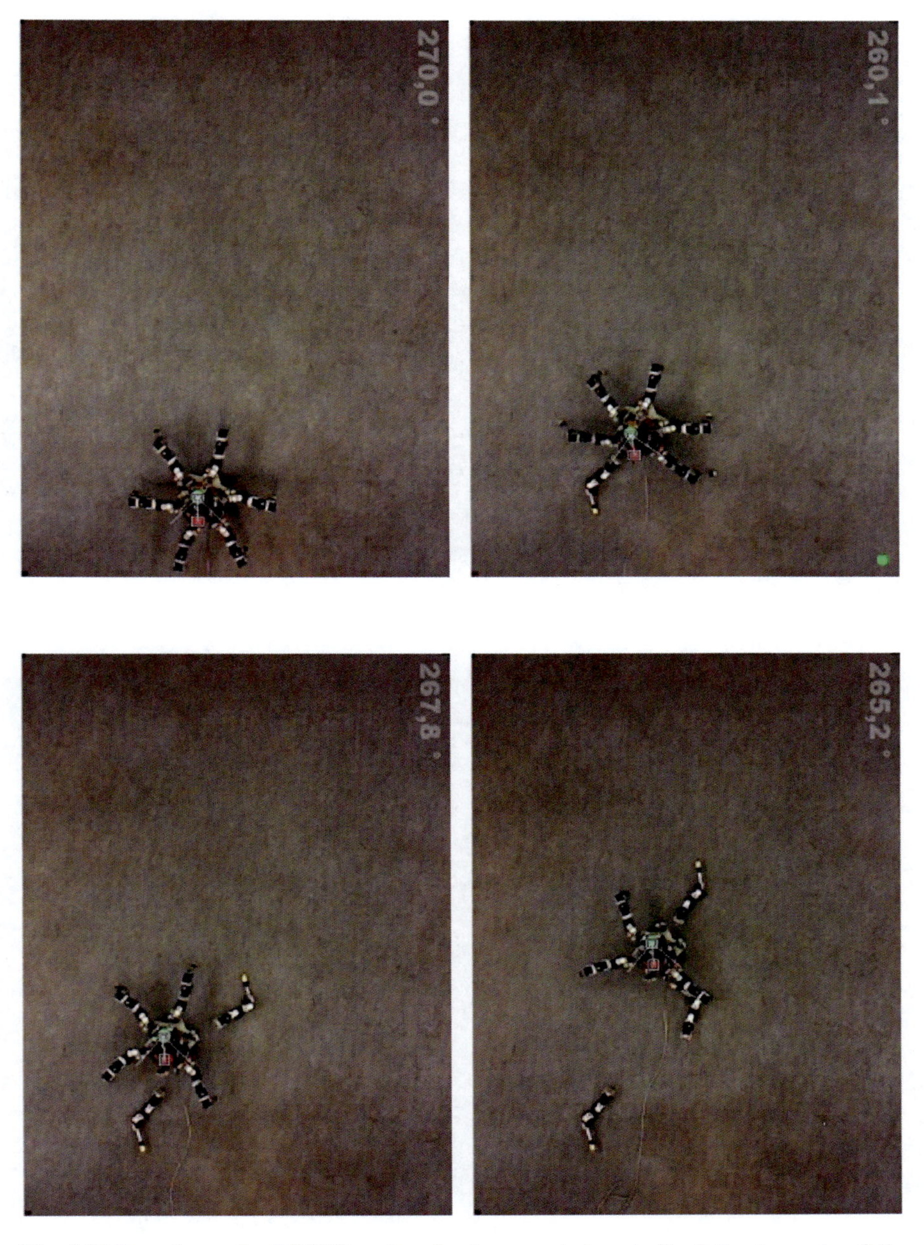

Fig. 8.18 Experiment 3 - OSCAR performing leg amputations in the following order: fully functional, leg 5 amputated, leg 1 amputated, leg 2 amputated - from left to right and from top to bottom.

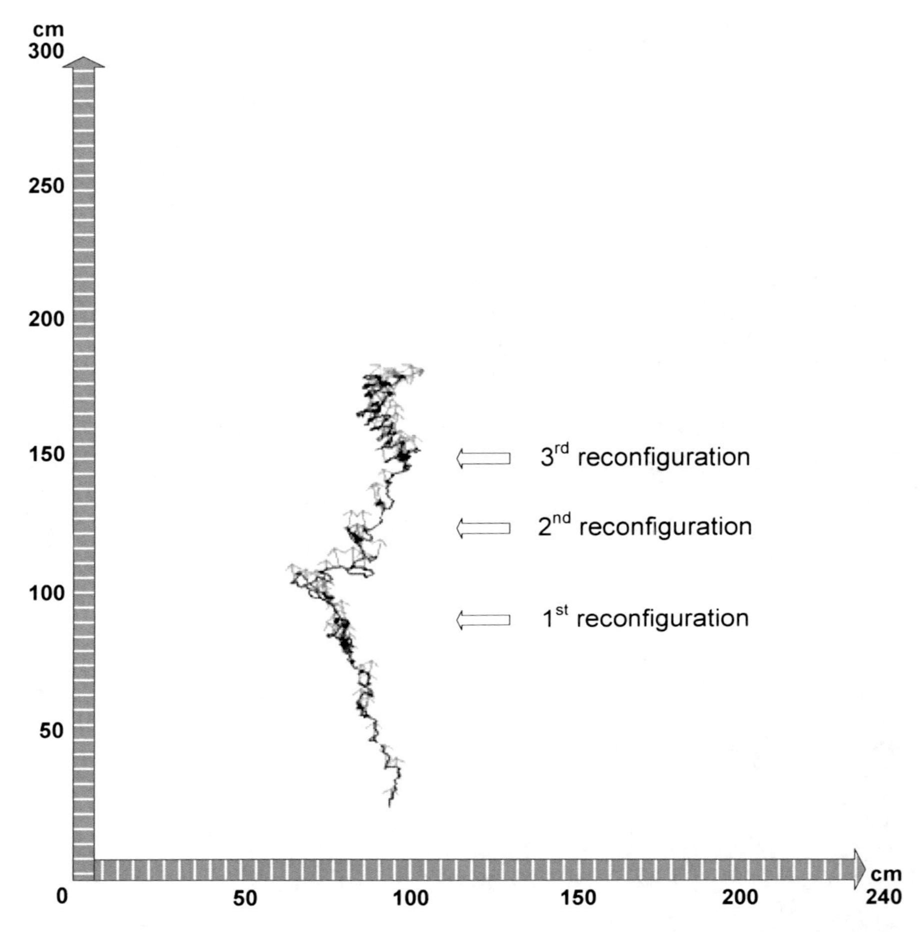

Fig. 8.19 Experiment 3 - Tracking the robot's heading while ejecting legs during its walking in the following order: 5, 1, 2.

Discussion for the tracking measurements done

On one hand it is nice to have a robotic system that exhibits emergent walking. On the other hand, this kind of pure emergent walking has a negative influence on the robot walking straight and its heading maintenance.

Despite this fact, it was still measured how the robot deviates from the straight path (keeping the course to 270°) while performing the leg amputations and walking with emergent gait. The results show that even when the robot has (virtual) malfunctions within its legs and performs legs amputations, it is still more or less capable of walking straight forward with slight turning in some cases (Figure 8.17). Although this deviation from course is present, we must take into account that the robot has amputated legs and that the deviation is still not that radical - like for example: the robot immediately walking in circles, or similar behavior.

One additional idea that might be used to avoid or minimize such deviation from the main course heading is to couple the emergent behavior with some other behaviors like going right or left, which in that case will somehow intervene with the emergent walking gait to keep the robot on its course.

8.5 Summary for the S.I.R.R. - Biologically Inspired Robot Reconfiguration Approach

In this chapter a biologically inspired Swarm Intelligence for Robot Reconfiguration (S.I.R.R.) approach was introduced and explained in detail. The S.I.R.R. approach was used for performing in-situ robot self-reconfiguration (using the introduced patent pending R-LEGAM mechanism). The S.I.R.R method is used for spatial distribution of the robot's legs when a reconfiguration is performed on the robot. Thus the robot achieves a stable spatial configuration after performing self-reconfiguration even when one or more legs malfunction and get amputated from the robot's body.

Through several experimental test cases and robot tracking it was demonstrated how the hexapod robot OSCAR-X continues with its walking despite the (virtual) anomalies that occur within its legs after amputating the malfunctioning legs and performing self-reconfiguration. In these experiments tracking measurements were also done on the robot's heading while performing its leg amputations and self-reconfigurations.

The presented results from experiments on *self-reconfiguration* look promising. Future work can consider additional research on integrating self-reconfiguration with walking robot's high-level behaviors, this aiming to improve the robot's heading after reconfiguration is preformed. One example could be improving of the curve walking [ELL07], although other different walking scenarios can be considered as well. Additional work can be also done on improving the robustness and generic usefulness of the presented self-reconfiguration approach and its potential application for other types of robots.

Chapter 9
Conclusion and Outlook

The biologically inspired approaches presented in this book are direct contributions to bionics research and organic computing for the domain of fault-tolerant robots.

Several newly introduced biologically inspired approaches have been explained in detail, and their practical usefulness proved through numerous experiments done on walking robot demonstrators. The approaches have demonstrated several important properties for the next generation of autonomous robots:

self-organization, self-adaptation, self-synchronization, self-stabilization, self-reconfiguration.

Self-organizing robot control architecture

In the 4th chapter, several author's ideas have been introduced concerning enhanced biologically inspired "stem" type ORCA architecture. Notions about such "stem" based control architecture are interesting for understanding self-organizing robot control architectures. The idea behind such *self-organizing* robot control architecture for a complex robotic system is that the burden for development of such robot control architecture can be drastically reduced. Such control architecture will also be suitable for approaches where robot *self-reconfiguration* is needed, so the robot would be able to self-reconfigure the components in its control architecture after some system's malfunction without human intervention, and continue with its mission despite some malfunctions it has overcome.

Future work will put more effort toward practical testing of the introduced notions for such self-organizing robot control architecture in real case situations and experiments.

Self-organizing emergent robot walking gait with distributed pressure on the robot's feet

In the 5th chapter, a concept for *self-organizing* emergent robot walking gait with distributed pressure on the robot's feet was introduced and explained in detail. Also, results from real tests done on the hexapod robot OSCAR-2 were presented.

B. Jakimovski: Biologically Inspired Approaches, COSMOS 14, pp. 175–178.
springerlink.com

This approach is useful for joint-leg walking robots, so they can traverse terrain by using self-organizing walking gait with distributed pressure on their feet. Distributed pressure on the robot's feet ensures that the robot is always statically and dynamically stable during its walking (over even or unstructured terrain). This pressure distribution also decreases the probability that the robot will tip over when walking on unstructured terrain.

Future work may involve combining self-organizing robot walking gait by distributed pressure on the robot's feet with standard robot walking behaviors.

Firefly inspired self-synchronization of a robot's walking gait

Another approach introduced in the 5th chapter is the biologically firefly inspired *self-synchronization* of the robot's walking gait. Here details are introduced on how to achieve walking gait self-synchronization by multi-legged walking robots. Results from simulation and real case experiments with this approach are also presented.

The simulations and real case experiments are related to three scenarios of walking gait self-synchronization: prolongation, shortening, and a combination prolongation and shortening of the robot's swing and stance phases using firefly inspired synchronization. Simulation experiments have been further verified through real case experiments done on a hexapod robot, including visual tracking of the robot. The results from tests show that the firefly inspired self-synchronization approach can be useful for self-synchronization of the robot's walking gait. Due to its generic property, it can also be applied to different kinds of multi-legged walking robots. Applying this concept will directly increase the autonomity of the robot while performing its mission tasks and at the same time decrease the human effort for programming such a robotic system with alternating walking gaits.

Future work will involve coupling the self-synchronization of the robot's walking gait with several robot behaviors and coordinating them so the robot can reliably hold its heading during such alternating robot's walking gaits with self-synchronization.

Self-stabilization of humanoid robot walking gait

In chapter 6, the SelSta approach (a biologically inspired Symbiosis) for *self-stabilizing* walking gait of a humanoid robot is introduced and explained in detail. The usefulness of the SelSta approach for the domain of walking humanoid robots is that the humanoid robots using the SelSta approach are able to optimize their walking gait for a particular flat surface. The self-stabilizing walking gait is not only optimized for increasing the stability of the robot during its walking, but also optimized for lowering the energy consumption while walking.

The results from experiments done on the humanoid robot S2-HuRo are presented in 4 different sections related to humanoid robot self-stabilizing walking over 4 different types of flat surfaces. They show that by using the SelSta approach the humanoid robot can find the optimal walking gait for a particular surface. These SelSta found walking parameters also introduce maximum stability of the robot during its walking and at the same time decrease the energy consumption to minimum levels. SelSta especially demonstrates its usefulness by finding the optimal robot walking gait for a particular surface in just 11-16 minutes, which in drastically lower than the 10-14 hours spent to manually configure quasi-optimal humanoid robot walking.

Future work will involve enhancing the Selsta approach to be useful for self-stabilizing the walking gait of a humanoid robot on unstructured terrains.

Self-adapting, biologically inspired RADE approach for anomaly detection

In chapter 7 a biologically inspired Artificial Immune System (AIS) based *self-adapting* Robot Anomaly Detection Engine (RADE) approach was introduced and explained in detail. The aim of this approach was to introduce a self-adapting anomaly detection method for robotic systems. The self-adaptation feature of RADE was explained in detail and its practical usefulness demonstrated through several test cases with hexapod robots. The results from the experiments show that RADE indeed is capable of adapting to a situation and generate an appropriate anomaly level which characterizes if the robot and its parts are in functional or non-functional condition. Due to the dynamics within RADE, the method quickly adapts to the situation and the anomaly level is generated in a very short time compared to a manually tuned generic fuzzy logic rule database for fault detection. Additionally, the RADE approach also reduces the effort for completely manually predefining the fault-detection fuzzy logic rule database (by fuzzy logic rule based fault detection methods), since the dynamics of RADE "guides" the anomaly level generation.

Future work can involve expansion of RADE by enabling the online learning feature so that the robot learns the normal and anormal conditions by itself while walking.

Self-reconfiguration of a hexapod robot based on a biological inspiration - swarm intelligence

In the 8th chapter, a novel approach for *self-reconfiguration* of a hexapod robot system was explained. The S.I.R.R. approach is based on the biological inspiration of swarm intelligence and it is introduced for the domain of multi-legged fault tolerant robots performing reconfiguration of their posture. The importance and main benefit of using the S.I.R.R. approach is that the robot realizes that some of its parts/legs have malfunctioned and performs reconfiguration of the posture of its still functional legs. S.I.R.R. spatially distributes the still functional legs after the malfunctioning legs have been amputated using the R-LEGAM mechanism.

The simulation results have shown how the hexapod robot would perform the reconfiguration with and without the S.I.R.R. approach. This gives a direct comparison and presents the usefulness of this approach for robot weight distribution and increase of stability for the walking robot after the robot has performed self-reconfiguration. The real case tests with OSCAR-2 using simulated leg amputations have shown how the hexapod robot reconfigures the legs and distributes the weight after the reconfiguration has been performed. The results from self-reconfiguration experiments done with the S.I.R.R. approach and real leg amputations on the hexapod robot OSCAR-X demonstrate the practical usefulness of this approach for reconfiguration purposes of multi-leg fault tolerant walking robots.

The results from tracking experiments have shown that future work may involve combining the self-reconfiguration approach with other higher level behaviors of the robot to maintain a proper heading even after the robot has amputated malfunctioning legs and performed reconfiguration.

Self-x features - all together working in a synergy

The self-x approaches mentioned in this work can be combined and work in synergy towards realizing robust and fault tolerant walking and hybrid robots. For example a multi-legged robot may use the self-organizing "stem" type control architecture ORCA presented in this work to properly map the parts that belong to the system during the start-up of the robot or after reconfiguration.

It will walk using the self-organizing emergent walking gait over variety of terrains and in case of hitting some obstacle it may use the firefly inspired self-synchronization method to re-synchronize its walking gait.

During its walking it can use the self-adapting robot anomaly detection engine (RADE) in order to check if the system is properly working, and in case of faults on some of its legs it may amputate the malfunctioned legs using the R-LEGAM amputation mechanism. Then after it has performed self-reconfiguration using the S.I.R.R. approach, and reconfiguration of the posture of its legs it will continue with its mission. During the reconfiguration the self-organizing control architecture will update the constituent elements of the system, and the firefly self-synchronization will aid the robot to re-synchronize its walking gait after the reconfiguration and to continue with its mission.

References

[AgP07] Aghili, E., Parsa, K.: Configuration Control and Recalibration of a New Reconfigurable Robot. In: IEEE International Conference on Robotics and Automation, pp. 4077–4083 (2007)

[AhP00] Amadjian, V., Paracer, S.: Symbiosis: an introduction to biological associations. Oxford University Press, Oxford (2000)

[ALD10] `http://www.aldebaran-robotics.com/en/`

[AMK01] Altendorfer, R., Moore, E.Z., Komsuoglu, H., Buehler, M., Brown, H., McMordie, D., Saranli, U., Full, R., Koditschek, D.E.: RHex: A Biologically Inspired Hexapod Runner. Autonomous Robots 11, 207–213 (2001)

[AnD90] Anderson, T.L., Donath, M.: Autonomous robots and emergent behavior: a set of primitive behaviors for mobile robot control. In: IEEE International Workshop on Intelligent Robots and Systems 1990, IROS 1990. Proceedings Towards a New Frontier of Applications, vol. 2, pp. 723–730 (1990)

[Ark87] Arkin, R.C.: Motor Schema Based Navigation for a Mobile Robot. In: Proceedings of the IEEE Conference on Robotics and Automation, pp. 264–271 (1987)

[AWY99] Akimoto, K., Watanabe, S., Yano, M.: An insect robot controlled by the emergence of gait patterns. Artificial Life and Robotics 3, 102–105 (1999)

[BaK06] Bartsch, S., Kirchner, F.: Robust control of a humanoid robot using a bio-inspired approach based on central pattern generators, reflexes, and proprioceptive feedback. In: IEEE International Conference on Robotics and Biomimetics, pp. 1547–1552 (2006)

[BaK06] Babaoglu, O., Binci, T., Jelasity, M., Montresor, A.: Firefly-inspired Heartbeat Synchronization in Overlay Networks. In: First International Conference on Self-Adaptive and Self-Organizing Systems, SASO 2007, pp. 77–86 (2007)

[BeW89] Beni, G., Wang, J.: Swarm Intelligence. In: Proc. 7th Ann. Meeting of the Robotics Society of Japan, pp. 425–428. RSJ Press (1989)

[BFH03] Borst, C., Fischer, M., Haidacher, S., Liu, H., Hirzinger, G.: DLR-Hand II: Experiments and Experiences with Anthropomorphic Hand. In: Proceedings of the 2003 IEEE Conference on Robotics & Automation, Taipei, Taiwan, pp. 702–707 (2003)

[BDT99] Bonabeau, E., Dorigo, M., Theraulaz, G.: Swarm Intelligence: From Natural to Artificial System. Oxford University Press, New York (1999)

[Bio10] `http://en.wikipedia.org/wiki/Bionics` (February 2010)

[BMM05] Brockmann, W., Maehle, E., Mösch, F.: Organic Fault-Tolerant Control Architecture for Robotic Applications. In: 4th IARP/IEEE-RAS/EURON Workshop on Dependable Robots in Human Environments, Nagoya, Japan (2005)

[BMT63] Becker, A.J., McCulloch, E.A., Till, J.E.: Cytological demonstration of the clonal nature of spleen colonies derived from transplanted mouse marrow cells. Nature 197, 452–454 (1963)

[BRO86] Brooks, R.: A robust layered control system for a mobile robot. IEEE Journal of Robotics and Automation(legacy, pre-1988) 2(1), 14–23 (2008)

[BuB76] Buck, J., Buck, E.: Synchronous fireflies. Scientific American 234, 74–85 (1976)

[Buc37] Buck, J.: Studies on firefly: I. The effects of light and other agents on flashing in Photinus pyralis, with special reference and diurnal rhythm. Physiological Zoology, 45–58 (1937)

[Buc88] Buck, J.: Synchronous rhythmic flashing of fireflies. Quart. Rev. Biol. 63, 265–289 (1988)

[BuM67] Buonomicini, M., Magni, F.: Nervous control flashing in the firefly Luciola italica. L. Archives italiennes de Biologie, 323–338 (1967)

[Bot95] Bottani, S.: Pulse-Coupled Relaxation Oscillators: From Biological Synchronization to Self Organized Criticality. Phy. Rev. Lett. 74(21) (1995)

[CaD03] Cao, Y., Dasgupta, D.: An Immunogenetic Approach in Chemical Spectrum Recognition. In: Ghosh, T. (ed.) Advances in Evolutionary Computing, ch. 36 (2003)

[CaS78] Case, J.F., Strause, L.G.: Neurally controlled luminescent systems. In: Herring, P.J. (ed.) Bioluminiscence in Action, pp. 331–366 (1978)

[CDF01] Camazine, S., Deneubourg, J., Franks, N.R., Sneyd, J., Theraulaz, G., Bonabeau, E.: Self Organization in Biological Systems, p. 154. Princeton University Press, Princeton (2001)

[CJT03] Canham, R., Jackson, A.H., Tyrrell, A.: Robot Error Detection Using an Artificial Immune System. In: Proceedings of the 2003 NASA/DoD Conference on Evolvable Hardware (2003)

[CMN04] Carlson, J., Murphy, R.R., Nelson, A.: Follow-up analysis of mobile robot failures. In: Proceedings of IEEE International Conference on Robotics and Automation, ICRA 2004, vol. 5, pp. 4987–4994 (2004)

[CMZ08] Calderon, C.A.A., Mohan, R.E., Zhou, C.: Virtual-RE: A Humanoid Robotic Soccer Simulator. In: International Conference on Cyberworlds, pp. 561–566 (2008)

[CoB99] Cohen, A.H., Boothe, D.L.: Sensorimotor interactions during locomotion: principles derived from biological systems. Autonomous Robots 7-3, 239–245 (1999)

[COB08] Christensen, A.L., O'Grady, R., Birattari, M., Dorigo, M.: Fault detection in autonomous robots based on fault injection and learning. Journal Autonomous Robots 24(1), 49–67 (2008)

[Con92] Connell, J.H.: SSS: A Hybrid Architecture Applied to Robot Navigation. In: Proceedings of IEEE International Conference on Robotics and Automation, vol. 3, pp. 2719–2724 (1992)

[Cru76] Cruse, H.: On the function of the legs in the free walking stick insect Carausius morosus. Journal of Comparative Physiology 112, 235–262 (1976)

[Cyb09] http://www.cyberbotics.com/products/webots/

[CZT07] Cuevas, E., Zaldivar, D., Tapia, E., Rojas, R.: An Incremental Fuzzy Algorithm for the Balance of Humanoid Robots. Humanoid Robots: New Developments (2007)

[DeM05] De Luca, A., Mattone, R.: An identification scheme for robot actuator faults. In: Proc. IEEE/RSJ Int. Conf. on Intelligent Robots and Systems, pp. 1127–1131 (2005)

[DeV02] De Castro, L.N., Von Zuben, F.J.: Learning and optimization using the clonal selection principle. IEEE Transactions on Evolutionary Computation 6(3), 239–251 (2002)

[DFG04] German Science Foundation (DFG) Priority Program SPP 1183.Organic Computing (2004), http://www.organic-computing.de/spp

[DeJ02] De Castro, L.N., Jonathan, T.: Artificial Immune Systems: A New Computational Intelligence Approach, pp. 57–58 (2002) ISBN 1852335947

[DoD99] Dorigo, M., Di Caro, G.: The Ant Colony Optimization Meta-Heuristic. In: Corne, D., Dorigo, M., Glover, F. (eds.) New Ideas in Optimization, pp. 11–32. McGraw-Hill, New York (1999)

[DoS04] Dorigo, M., Stützle, T.: Ant Colony Optimization. MIT Press, Cambridge (2004)

[DWX08] Deng, X., Wang, J., Xiang, Z.: The Simulation Analysis of Humanoid Robot Based on Dynamics and Kinematics. In: Xiong, C.-H., Liu, H., Huang, Y., Xiong, Y.L. (eds.) ICIRA 2008. LNCS (LNAI), vol. 5314, pp. 93–100. Springer, Heidelberg (2008)

[EHS06] Everist, J., Hou, F., Shen, W.: Transformation of Control in Congruent Self-Reconfigurable Robot Topologies. In: IEEE/RSJ International Conference on Intelligent Robots and Systems, pp. 612–618 (2006)

[ELL07] El Sayed Auf, A., Larionova, S., Litza, M., Mösch, F., Jakimovski, B., Maehle, E.: Ein Organic Computing Ansatz zur Steuerung einer sechsbeinigen Laufmaschine. In: AMS, pp. 233–239. Springer, Heidelberg (2007)

[EML06] El Sayed Auf, A., Mösch, F., Litza, M.: How the Six-Legged Walking Machine OSCAR Handles Leg Amputations. From Animals to Animals 9 (2006)

[EPU09] http://www.e-puck.org

[FaC93] Ferrell, C.: Robust Agent Control of an Autonomous Robot with Many Sensors and Actuators. MIT Artificial Intelligence Lab Technical Report 1443 (1993)

[FPA94] Forrest, S., Perelson, A.S., Allen, L., Cherukuri, R.: Self-Nonself Discrimination in a Computer. In: Proceedings of the 1994 IEEE Symposium on Research in Security and Privacy. IEEE Computer Society Press, Los Alamitos (1994)

[FPV08] Friedmann, M., Petersen, K., von Stryk, O.: Tailored Real-Time Simulation for Teams of Humanoid Robots. In: Visser, U., Ribeiro, F., Ohashi, T., Dellaert, F. (eds.) RoboCup 2007: Robot Soccer World Cup XI. LNCS (LNAI), vol. 5001, pp. 425–432. Springer, Heidelberg (2008)

[Gen08] Jakimovski, B.: Generic robot architecture – Internal project. Institute of Computer Engineering, University Lübeck, Germany (2008)

[GKS04] Gudmundsson, J., Kreveld, M.V., Speckmann, B.: Efficient detection of motion patterns in spatiotemporal data sets. In: GIS 2004: Proceedings of the 12th Annual ACM International Workshop on Geographic Information Systems (2004)

[Gol10] Goldstein, J.: Emergence as a Construct: History and Issues. Emergence 11, 49–72 (1999)

[Gri81] Grillner, S.: Control of locomotion in bipeds, tetrapods and fish. In: Handbook of Physiology II, pp. 1179–1236. American Physiol. Society (1981)

[GUM09] http://www.gumstix.com

[GWH09] Gorner, M., Wimbock, T., Hirzinger, G.: The DLR Crawler: evaluation of gaits and control of an actively compliant six-legged walking robot. Industrial Robot: An International Journal 36(4), 344–351 (2009)

[HaH06] Hancher, M.D., Hornby, G.S.: A Modular Robotic System with Applications to Space Exploration. In: 2nd IEEE International Conference on Space Mission Challenges for Information Technology (SMC-IT 2006), pp. 125–132 (2006)

[HAN10] http://hansonrobotics.wordpress.com/about/

[HCB71] Hanson, F.E., Case, J.F., Buck, E., Buck, J.: Synchrony and flash entrainment in a New Guinea firefly. Science 174, 161–164 (1971)

[HIT06] http://www.hitecrobotics.com

[HLC06] Hyun Yool, K., Lee, Y., Choi, H., Hyun Yool, B., Hwan Kim, D.: Swarm Robotics: Self Assembly, Physical Configuration, and Its Control. In: SICE-ICASE International Joint Conference, pp. 4276–4279 (2006)

[HON07] http://asimo.honda.com/

[HUB03] http://hubolab.kaist.ac.kr/index.php

[HuE02] Hu, X., Eberhart, R.C.: Solving constrained nonlinear optimization problems with particle swarm optimization. In: Proceedings of the Sixth World Multiconference on Systemics, Cybernetics and Informatics (2002)

[IBM01] [IBM Research: Autonomic Computing, http://www.ibm.com/research/autonomic

[ITG09] Koo, I.M., Kang, T.H., Vo, G.L., Trong, T.D., Song, Y.K., Choi, H.R.: Biologically inspired control of quadruped walking robot. International Journal of Control, Automation and Systems 7, 577–584 (2009)

[IYS06] Inagaki, S., Yuasa, H., Suzuki, T., Arai, T.: Wave CPG model for autonomous decentralized multilegged robot: Gait generation and walking speed control. Robotics and Autonomous Systems (RAS) 54(2), 118–126 (2006)

[Jak09] Jakimovski, B.: Patent pending Robot Leg Amputation Mechanism (R-LEGAM) for joint leg walking robots. DPMA-Az: 10 2009 006 934.8

[JaM08] Jakimovski, B., Maehle, E.: Artificial immune system based robot anomaly detection engine for fault tolerant robots. In: Rong, C., Jaatun, M.G., Sandnes, F.E., Yang, L.T., Ma, J. (eds.) ATC 2008. LNCS, vol. 5060, pp. 177–190. Springer, Heidelberg (2008)

[JaM10] Jakimovski, B., Maehle, E.: In situ self-reconfiguration of hexapod robot OSCAR using biologically inspired approaches. In: Miripour (ed.) Climbing and Walking Robots INTECH, pp. 311–332 (2010) ISBN: 978-953- 307-030-8

[JKH10] Jakimovski, B., Kotke, M., Hörenz, M., Maehle, E.: SelSta - A Biologically Inspired Approach for Self-Stabilizing Humanoid Robot Walking. In: Distributed, Parallel and Biologically Inspired Systems, IFIP Advances in Information and Communication Technology, vol. 329, pp. 302–313 (2010)

[JMM09] Jakimovski, B., Meyer, B., Maehle, E.: Self-reconfiguring hexapod robot OSCAR using organically inspired approaches and innovative robot leg amputation mechanism. In: International Conference on Automation, Robotics and Control Systems, ARCS-2009, USA (2009)

[JMM10] Jakimovski, B., Meyer, B., Maehle, E.: Firefly flashing synchronization as inspiration for self-synchronization of walking robot gait patterns using a decentralized robot control architecture. In: Müller-Schloer, C., Karl, W., Yehia, S. (eds.) ARCS 2010. LNCS, vol. 5974, pp. 61–72. Springer, Heidelberg (2010)

[JMM08] Jakimovski, B., Meyer, B., Maehle, E.: Swarm Intelligence for Self-Reconfiguring Walking Robot. In: IEEE Swarm Intelligence Symposium, USA, St. Louis, Missouri, September 21-23 (2008)

[KCC06] Kim, S., Clark, J.E., Cutkosky, M.R.: iSprawl: Design and Tuning for High-speed Autonomous Open-loop Running. International Journal of Robotics Research 25(9), 903–912 (2006)

[KKR06] Kim, H.K., Kwon, W., Roh, K.S.: Biologically Inspired Energy Efficient Walking for Biped Robots. In: IEEE International Conference on Robotics and Biomimetics, pp. 630–635 (2006)

[KKS06] Kratz, R., Klug, S., Stelzer, M., von Stryk, O.: Biologically Inspired Reflex Based Stabilization Control of a Humanoid Robot with Artificial SMA Muscles. In: IEEE International Conference on Robotics and Biomimetics, pp. 1089–1094 (2006)

[Kot09] Kotke, M.: Biologisch inspiriertes Verfahren für die Selbststabilisierung eines humanoiden Roboters. Studienarbeit (2009)

[KSW05] Kopacek, P., Schierer, E., Wuerzl, M.: A Controller Network for a Humanoid Robot. In: Moreno Díaz, R., Pichler, F., Quesada Arencibia, A. (eds.) EUROCAST 2005. LNCS, vol. 3643, pp. 584–589. Springer, Heidelberg (2005)

[MDD04] McIntyre, M., Dixon, W., Dawson, D., Walker, I.: Fault detection and identification for robot manipulators. In: Proc. of 2004 IEEE Int. Conf. on Robotics and Automation, pp. 4981–4986 (2004)

[KeE95] Kennedy, J., Eberhart, R.C.: Particle swarm optimization. In: Proceedings of IEEE International Conference on Neural Networks, vol. IV, pp. 1942–1948 (1995)

[MEN08] Morimoto, J., Endo, G., Nakanishi, J., Cheng, G.: A Biologically Inspired Biped Locomotion Strategy for Humanoid Robots: Modulation of Sinusoidal Patterns by a Coupled Oscillator Model. IEEE Transactions on Robotics 24, 185–191 (2008)

[KeC03] Kephart, J.O., Chess, D.M.: The Vision of Autonomic Computing. Computer 36(1), 41–50 (2003)

[LHG06] Liu, J., Hu, H., Gu, D.: A hybrid control architecture for autonomous robotic fish. In: Proc. Int. Conf. Intelligent Robots and Systems, pp. 312–317 (2006)

[LiC05] Liu, H., Coghill, G.M.: A model-based approach to robot fault diagnosis. In: Applications and Innovations in Intelligent Systems XII, pp. 137–150 (2005)

[LiJ07] Liu, Y., Jiang, J.: Fault Diagnosis Method for Mobile Robots Using Multi-CMAC Neural Networks. In: IEEE International Conference on Automation and Logistics, pp. 903–907 (2007)

[Lip07] Lipson, H.: Evolutionary robotics: emergence of communication. Curr. Biol. 17, 330–332 (2007)

[LSY07] Lee, B., Stonier, D., Yong-Duk, K., Jeong-Ki, Y., Jong-Hwan Kim, K.: Modifiable walking pattern generation using real-time ZMP manipulation for humanoid robots. In: IEEE/RSJ International Conference on Intelligent Robots and Systems, IROS 2007, pp. 4221–4226 (2007)

[Lyn06] `http://www.lynxmotion.com/Category.aspx?CategoryID=92`

[LVO01] Lewandowski, S.M., Van Hook, D.J., O'Leary, G.C., Haines, J.W., Rossey, L.M.: SARA: Survivable Autonomic Response Architecture. In: Proc. DARPA Information Survivability Conference and Exposition II, vol. 1, pp. 77–88 (2001)

[Mag67] Magni, F.: Central and peripheral mechanisms in the modulation of flashing in the firefly Luciola italica. L. Archives italiennes de Biologie, 339–360 (1967)

[Mat87] Matsuoka, K.: Mechanisms of frequency and pattern control in the neural rhythm generators. Biological Cybernetics 56, 345–353 (1987)

[MAV09] Monteriu, A., Asthana, P., Valavanis, K.P., Longhi, S.: Real-Time Model-Based Fault Detection and Isolation for UGVs. Journal of Intelligent and Robotic Systems (JIRS), 425–439(2007)

[MDB07] Makarov, V.A., Del Rio, E., Bedia, M.G., Velarde, M.G., Ebeling, W.: Central pattern generator incorporating the actuator dynamics for a hexapod robot. Int. J. Applied Math. & Comp. Sci. 3(2), 97–102 (2007)

[MiV02] Michelan, R., Von Zuben, F.J.: Decentralized control system for autonomous navigation based on an evolved artificial immune network. In: Proceedings of the 2002 Congress on Evolutionary Computation, vol. 2, pp. 1021–1026 (2002)

[Msd09] `http://msdn.microsoft.com/en-us/robotics/default.aspx`

[Mül04] Müller-Schloer, C.: Organic Computing: On the Feasibility of Controlled Emergence. In: Proceedings of the 2nd IEEE/ACM/IFIP International Conference on Hardware/Software Codesign and System Synthesis, New York, USA, pp. 2–5 (September 2004)

[MuK07] Murata, S., Kurokawa, H.: Self-Reconfigurable Robot: Shape-Changing Cellular Robots Can Exceed Conventional Robot Flexibility. IEEE Robotics & Automation Magazine (2007)

[NaB07] Nami, M.R., Bertels, K.: A Survey of Autonomic Computing Systems. In: Proceedings of the Third International Conference on Autonomic and Autonomous, pp. 101–110 (2007)

[NAS08] Lessons in Implementing Bio-inspired Algorithms on Wireless Sensor Networks. In: NASA/ESA, Conference on Adaptive Hardware and Systems, pp. 271-276 (2008)

[NAS10] `http://robonaut.jsc.nasa.gov/about.asp`

[Nat07] `http://www.ni.com`

[Nat10] `http://www.n100best.org/list.html` (February 2010)

[NCR05] Nasraoui, O., Cardona, C., Rojas, C.: Using retrieval measures to assess similarity in mining dynamic web clickstreams. In: Proceeding of the Eleventh ACM SIGKDD International Conference on Knowledge Discovery in Data Mining KDD (2005)

[NFR06] Neal, M., Feyereisl, J., Rascuna, R., Wang, X.: Don't Touch Me, I'm Fine: Robot Autonomy Using an Artificial Innate Immune System. In: 5th International Conference on Artificial Immune Systems (2006)

[Ope09] `http://sourceforge.net/projects/opencvlibrary/`

[PaH09] Panagou, D., Herbert, T.: Modeling of a Hexapod Robot; Kinematic Equivalence to a Unicycle. UDME Technical Report (2009)

[PaV05] Pagnoni, A., Visconti, A.: An innate immune system for the protection of computer networks. In: Proceedings of The 4th International Symposium on Information And Communication Technologies. ACM International Conference Proceeding Series, vol. 92, pp. 63–68 (2005)

[Pla09] `http://playerstage.sourceforge.net/gazebo/gazebo.html`

[POP07] Prada, R., Otero, N., Paiva, A.: The user in the group: evaluating the effects of autonomous group dynamics. In: ACE 2007: Proceedings of the International Conference On Advances In Computer Entertainment Technology, pp. 25–32 (2007)

[Prz06] Przystalka, P.: Model-Based Fault Detection and Isolation Using Locally Recurrent Neural Networks. In: Rutkowski, L., Tadeusiewicz, R., Zadeh, L.A., Zurada, J.M. (eds.) ICAISC 2008. LNCS (LNAI), vol. 5097, pp. 123–134. Springer, Heidelberg (2008)

[PTP03] Poulton, J.W., Tell, S., Palmer, R.: Multiwire Differential Signaling. UNC-CH Department of Computer Science, Revision 1.1, pp. 1–3 (2003)

[QRY08] Qing, T., Rong, X., Yong, L., Jian, C.: HumRoboSim: An Autonomous Humanoid Robot Simulation System. In: 2008 International Conference on Cyberworlds, pp. 537–542 (2008)

[Rey87] Reynolds, C.: Flocks, herds, and schools: A distributed behavioral model. Comp. Graph. 21(4), 25–34 (1987)

[Rey88] Reynolds, C.: Not Bumping Into Things. In: Notes on obstacle Avoidance for the Course on Physically Based Modeling at SIGGRAPH 1988, Atlanta (1988)

[RMB06] Richter, U., Mnif, M., Branke, J., Müller-Schloer, C., Schmeck, H.: Towards a generic observer/controller architecture for organic computing. In: INFORMATIK 2006 – Informatik für Menschen. LNI,vol. P-63 pp. 112–119, GI-Edition (2006)

[Rob10] `http://www.robocup.org`

[RRT07] Rouff, C., Rash, J., Truszkowski, W.: Overcoming Robotic Failures through Autonomicity. In: Fourth IEEE International Workshop on Engineering of Autonomic and Autonomous Systems, EASe 2007. pp. 154–162 (2007)

[SaS02] Sathyanath, S., Sahin, F.: AISIMAM – An Artificial Immune System Based Intelligent Multi Agent Model and its Application to a Mine Detection Problem. In: 1st International Conference on Artificial Immune Systems (2002)

[Sch05] Schmeck, H.: Organic Computing – A New Vision for Distributed Embedded Systems. In: Proceedings of the Eighth IEEE International Symposium on Object-Oriented Real-Time Distributed Computing (ISORC 2005), pp. 201–203. IEEE, IEEE Computer Society, Los Alamitos (2005)

[SCM06] Spenko, M., Cutkosky, M., Majidi, C., Fearing, R., Groff, R., Autumn, K.: Foot design and integration for bioinspired climbing robots. In: Proc. SPIE Defense & Security Symposium. Unmanned Systems Technology, Orlando (2006)

[SGF06] Saunders, A., Goldman, D.I., Full, R.J., Buehler, M.: The RiSE climbing robot: body and leg design. In: Proceedings of the SPIE, vol. 6230 (2006)

[SHR06] Sterritt, R., Hinchey, M., Rouff, C., Rash, J., Truszkowski, W.: Sustainable and Autonomic Space Exploration Missions. In: 2nd IEEE International Conference on Space Mission Challenges for Information Technology (SMC-IT 2006), pp. 59–66 (2006)

[ShT07] Shkolnik, A., Tedrake, R.: Inverse kinematics for a point-foot quadruped robot with dynamic redundancy resolution. In: Proceedings of the 2007 IEEE International Conference on Robotics and Automation (2007)

[SiN05] Singh, C.T., Nair, S.B.: An Artificial Immune System for a MultiAgent Robotics System. Transactions of Engineering, Computing and Technology 6, 308–311 (2005)

[SpM79] Spirito, C., Mushrush, D.: Interlimb coordination during slow walking in the cockroach I. Effects of substrate alterations. Journal Exp. Biol. 78, 233–243 (1979)

[SMK06] Shimizu, M., Mori, T., Kawakatsu, T., Ishiguro, A.: An Adaptive Morphology Control of a Modular Robot. In: International Joint Conference on SICE-ICASE. pp. 4509–4514 (2006)

[SMT63] Siminovitch, L., McCulloch, E.A., Till, J.E.: The distribution of colony-forming cells among spleen colonies. Journal of Cellular and Comparative Physiology 62, 327–336 (1963)

[STE10] `http://stemcells.nih.gov/staticresources/info/basics/StemCellBasics.pdf` (March 2010)

[STW10] Steingrube, S., Timme, M., Wörgötter, F., Manoonpong, P.: Self-organized adaptation of a simple neural circuit enables complex robot behaviour. Nature Physics (January 2010)

[TOS09] `http://topio.tosy.com/about.shtml`

[TSB04] Thorstensen, B., Syversen, T., Bjørnvold, T., Walseth, T.: Electronic shepherd - a low-cost, low bandwidth, wireless network system. In: MobiSys 2004: Proceedings of the 2nd International Conference on Mobile Systems, Applications and Services (2004)

[TsL06] Tsay, T.I.J., Lai, C.H.: Design and Control of a Humanoid Robot. In: 2006 IEEE/RSJ International Conference on Intelligent Robots and Systems, vol. 9, pp. 2002–2007 (2006)

[TTK01] Tsuchiya, K., Tsujita, K., Kawakami, M., Aoi, S.: An Emergent Control of Gait Patterns of Legged Locomotion Robots. In: Proceedings of 4th IFAC Symposium on Intelligent Autonomous Vehicles, pp. 271–276 (2001)

[Usa09] `http://sourceforge.net/projects/usarsim`

[VBS90] Vukobratovic, M., Borovac, B., Surla, D., Stokic, D.: Biped Locomotion, Dynamics, Stability, Control and Application. Springer, Berlin (1990)

[VeS06] Verma, V., Simmons, R.: Scalable robot fault detection and identification. Robotics and Autonomous Systems 54(2), 184–191 (2006)

[VuJ69] Vukobratovic, M., Juricic, D.: Contribution to the synthesis of biped gait. IEEE Trans. Bio-Med. Eng. BME-16(1), 1–6 (1969)

[Vuk73] Vukobratovic, M.: How to control artificial anthropomorphic system. IEEE Trans. Syst. Man. Cyb. SMC-3(5), 497–507 (1973)

[WHK06] Wischmann, S., lse, M., Knabe, J., Pasemann, F.: Synchronization of internal neural rhythms in multi-robotic systems. Adaptive Behavior 14, 117–127 (2006)

[Win67] Winfree, A.T.: Biological rhythms and the behavior of populations of coupled oscillators. Journal of Insect Physiology 15, 597–610 (1967)

[Win06] Winfield, A.F.T.: Safety in numbers: fault-tolerance in robot swarms. International Journal of Modelling, Identification and Control 1, 30–37 (2006)

[XMW08] Xuan, L., Minglu, Z., Wei, L.: Methods to Modular Robot Design. In: Second International Symposium on Intelligent Information Technology Application, pp. 663–668 (2008)

[YaL03] Yang, D.C., Liu, L.: Kinematic Analysis of Humanoid Robot. Chinese J. of Mechanical Engineering 39, 70–74 (2003)

[YAL06] Yumaryanto, A.A., An, J., Lee, S.: Development of a Biologically-Inspired Mesoscale Robot. In: Yang, Q., Webb, G. (eds.) PRICAI 2006. LNCS (LNAI), vol. 4099, pp. 875–879. Springer, Heidelberg (2006)

[YuT08] Yu, J., Tirkkonen, O.: Self-Organized Synchronization in Wireless Network. In: Second IEEE International Conference on Self-Adaptive and Self-Organizing Systems, pp. 329–338 (2008)

[Zad65] Zadeh, L.A.: Fuzzy sets. Information and Control 8, 338–353 (1965)

[Zha06] Zhang, Y.: Autonomous Robot Failure Recognition Design using Multi-Objective Genetic Programming. In: Proceedings of the Fifth International Conference on Machine Learning and Cybernetics, pp. 4563–4568 (2006)

[ZHH98] Zheng, Z., Hu, G., Hu, B.: Phase Slips and Phase Synchronization of Coupled Oscillators. Physical Review Letters 81 (1998)

[ZMZ06] Zhuo-hua, D., Ming, F., Zi-xing, C., Jin-xia, Y.: An adaptive particle filter for mobile robot fault diagnosis. Journal of Central South University of Technology 13(6), 689–693 (2006)

A Appendix

A.1 Test Bed for Tracking the Robot OSCAR-X during the Experiments

The tracking of the robot's walking was done with a web camera and specially made tracking software utilizing the OpenCV library [Ope09] and its methods for blob tracking. There were also one red and one green circle placed on the robot (Figure A.1). The red circle was mounted on the back of the robot between legs 0 and 5 (Figure A.1, Figure A.3), the green circle on the front of the robot between legs 2 and 3 (Figure A.1, Figure A.3). The colored circles are recognized by the tracking software and a vector pointing from the red circle to the green circle shows the direction in which the robot is walking.

The test setup is presented in Figure A.2. The web camera is positioned about 4.5 meters above the terrain on which the robot is walking. The camera is connected to the computer (PC) where the tracking, recording of video, and logging of the orientation of the robot is done by the specially prepared tracking software.

Fig. A.1 Setup of the robot OSCAR-X with red and green indicators that aid in tracking with camera.

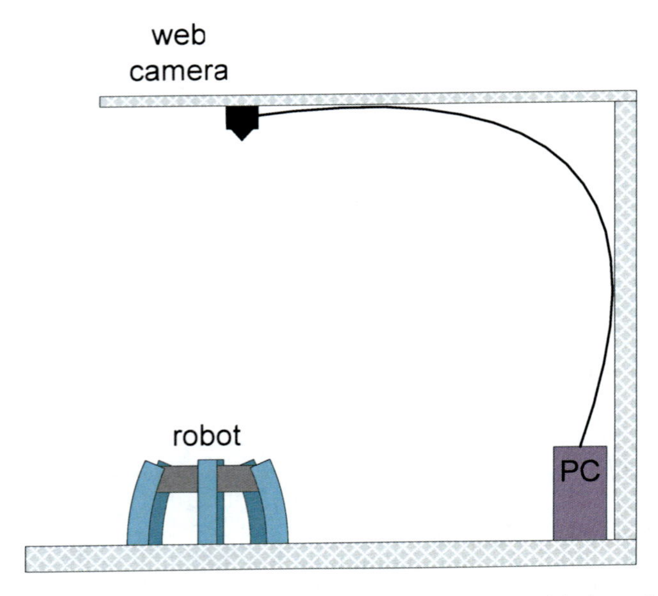

Fig. A.2 Schematic of the tracking setup for robot's walking with the web camera and a specially developed tracking software running on a computer (PC).

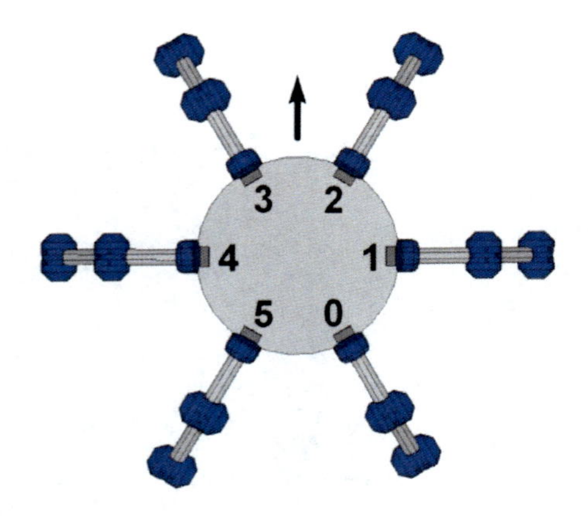

Fig. A.3 Model of a hexapod robot with its legs numbered.

List of Figures

Keywords

Six legged robot, walking robot, hexapod robot OSCAR, fault tolerant walking robot, decentralized robot control architecture, robot anomaly detection, robot fault detection, artificial immune system, self-adapting, robot anomaly detection engine, RADE, robot leg amputation, robot leg ejection mechanism, R-LEGAM, S.I.R.R., swarm intelligence for robot reconfiguration, self-reconfiguration, self-reconfiguring hexapod robot, dynamically prolongation and shortening of robot's walking gait patterns, emergent robot gait synchronization, firefly synchronization, humanoid robot, self-synchronization, self stabilizing humanoid robot, S2-HuRo, biologically inspired approach, humanoid robot walking optimization, symbiosis, SelSta, symbiosis score, self-stabilization, self-optimization.

Glossary

AIS	Artificial Immune System
BCU	Basic Control Unit
CPG	Central Pattern Generator
DAQ	Data acquisition
DOF	Degrees Of Freedom
OCU	Organic Control Unit
RADE	Robot Anomaly Detection Engine
S.I.R.R.	Swarm Intelligence for Robot Reconfiguration
OSCAR	Organic Self Configuring and Adapting Robot
ORCA	Organic Robot Control Architecture
PC	Personal Computer
S2-HuRo	Self Stabilizing Humanoid Robot
SelSta	Self-Stabilization (Approach)
SymbScore	Symbiosis Score
ZMP	Zero Moment Pole

Cognitive Systems Monographs

Edited by R. Dillmann, Y. Nakamura, S. Schaal and D. Vernon

Vol. 1: Arena, P.; Patanè, L. (Eds.)
Spatial Temporal Patterns for
Action-Oriented Perception
in Roving Robots
425 p. 2009 [978-3-540-88463-7]

Vol. 2: Ivancevic, T.T.; Jovanovic, B.;
Djukic, S.; Djukic, M.; Markovic, S.
Complex Sports Biodynamics
326 p. 2009 [978-3-540-89970-9]

Vol. 3: Magnani, L.
Abductive Cognition
534 p. 2009 [978-3-642-03630-9]

Vol. 4: Azad, P.
Visual Perception for Manipulation
and Imitation in Humanoid Robots
270 p. 2009 [978-3-642-04228-7]

Vol. 5: de Aguiar, E.
Animation and Performance Capture
Using Digitized Models
168 p. 2010 [978-3-642-10315-5]

Vol. 6: Ritter, H.; Sagerer, G.;
Dillmann, R.; Buss, M.:
Human Centered Robot Systems
216 p. 2009 [978-3-642-10402-2]

Vol. 7: Levi, P.; Kernbach, S. (Eds.):
Symbiotic Multi-Robot Organisms
467 p. 2010 [978-3-642-11691-9]

Vol. 8: Christensen, H.I.;
Kruijff, G.-J.M.; Wyatt, J.L. (Eds.):
Cognitive Systems
491 p. 2010 [978-3-642-11693-3]

Vol. 9: Hamann, H.:
Space-Time Continuous Models of Swarm
Robotic Systems
147 p. 2010 [978-3-642-13376-3]

Vol. 10: Allerkamp, D.:
Tactile Perception of Textiles in a
Virtual-Reality System
120 p. 2010 [978-3-642-13973-4]

Vol. 11: Vernon, D.; von Hofsten, C.; Fadiga, L.:
A Roadmap for Cognitive Development
in Humanoid Robots
227 p. 2010 [978-3-642-16903-8]

Vol. 12: Ivancevic, T.T.; Jovanovic, B.;
Jovanovic, S.; Djukic, M.; Djukic, N.;
Lukman, A.:
Paradigm Shift for Future Tennis
375 p. 2011 [978-3-642-17094-2]

Vol. 13: Bardone, E.:
Seeking Chances
168 p. 2011 [978-3-642-19632-4]

Vol. 14: Jakimovski, B.:
Biologically Inspired Approaches for
Locomotion, Anomaly Detection and
Reconfiguration for Walking Robots
201 p. 2011 [978-3-642-22504-8]